AN INTRODUCTION TO
COMPACT LIE GROUPS

SERIES IN PURE MATHEMATICS

Editor: C C Hsiung
Associate Editors: S S Chern, S Kobayashi, I Satake, Y-T Siu, W-T Wu
 and M Yamaguti

Part I. Monographs and Textbooks

Part II. Lecture Notes

Series in Pure Mathematics — Volume 13

INTRODUCTION TO COMPACT LIE GROUPS

Howard D Fegan
Department of Mathematics and Statistics
University of New Mexico
USA

World Scientific
Singapore • New Jersey • London • Hong Kong

Published by

World Scientific Publishing Co. Pte. Ltd.

P O Box 128, Farrer Road, Singapore 9128

USA office: Suite 1B, 1060 Main Street, River Edge, NJ 07661

UK office: 73 Lynton Mead, Totteridge, London N20 8DH

First published 1991
First reprint 1998

AN INTRODUCTION TO COMPACT LIE GROUPS

ISBN   981-02-0702-6
ISBN   981-02-3686-7 (pbk)

Printed in Singapore by JCS Office Services & Supplies Pte Ltd

For Ann

Contents

Preface

This book grew out of a seminar held at the University of New Mexico. Its aim is to provide the reader with the tools to do analysis on Lie groups. Of course, much standard material is presented which can be used for many different purposes. It is intended to provide the beginning graduate student with access to the subject of compact Lie groups. However, the material should also be suitable for senior undergraduates who are willing to spend the time that the subject requires. The other segment of the intended audience is that of research workers in other fields.

The subject of compact Lie groups is a rather old subject which is enjoying renewed attention. Since it is fairly old there has been time for the results to mature. Thus, there is a good compromise between stating results in great generality, and so losing sight of the details, or conversely in becoming too deeply embedded in details. It is this maturity which makes the subject ideal as an introduction to advanced mathematics. The other reason for the renewal of interest is the wide area of applications. The most outstanding example of this comes from gauge theory in physics. Here the mathematical model of the physical situation involves a principal fibre bundle with a compact Lie group as fibre.

A key feature of the book is the mixture of abstract theory and concrete example. In fact, there was a general plan for each chapter: to start with, a discussion of the theory, and then illustrate it in the case of some of the classical matrix groups. Not every chapter naturally fits such a plan, and when this happens, the plan has been violated so that the mathematics can appear as naturally as possible. The most obvious exception is the first chapter. As a consequence of this procedure, we have given a detailed explicit description of the structure, representation theory, and differential geometry of the vast majority of compact Lie groups, which most readers are likely to encounter in practice.

Now, for some detailed comments on the contents. A Lie group is both a group and a manifold. In Chapter 1, we study the basic theory of manifolds. Due to the special nature of compact Lie groups, relatively little of the theory of manifolds is actually needed. The discussion in Chapter 1 is generally restricted to such theory as is necessary to study Lie groups and describe the context of this theory.

Chapters two, three, and four give the basic structure of compact
Lie groups. In Chapter 2, we give the basic definitions. In Chapter 3,
we see the Lie algebra come into play. Lie algebras are, generally,
easier to work with than Lie groups. The exponential map enables us
to pass between the group and algebra. In fact, on a suitably small
part of the Lie algebra, the exponential map provides us with an isomorphism with part of the group. Chapter 4 discusses the Campbell-
Baker-Hausdorff formula, which shows one major manifestation of the
nonabelian nature of some Lie groups.

The transition between studying the structure of compact Lie
groups and their representation theory is in chapters five and six.
In Chapter 5, we meet the adjoint representation, which is obtained
from letting a group act on itself. This is used in Chapter 6 when we
show that any two maximal tori are conjugate. This is a major result
about the structure of compact Lie groups, which in turn will be used
to describe the representations of such groups.

We are now at a point where we can study representation theory.
This is the subject of chapters seven, eight, and nine. In Chapter 7,
we give the basic definitions. It is an amazing fact that one can
explicitly describe all the representations of a compact Lie group.
The classification is by highest weights, and is given in Chapter 8.
There are some remarkable formulae due to Hermann Weyl, which
are established in Chapter 9.

The final three chapters are about analysis on a compact Lie group.
They all use representation theory quite heavily. In Chapter 10, we
describe the invariant differential operators on a compact Lie group.
Chapter 11 gives the Levi-Civita connection and the various curva-
tures. To illustrate the use of this, we study the trace of the heat
kernel in Chapter 12. This includes a proof of the "strange formula"
of Freudenthal and deVries.

There are two appendices. The first is about the tensor product and
explains some multilinear algebra. In the second, we define Clifford
algebras and give a construction of the spin groups.

Finally, one should note that each chapter finishes with a set of
exercises. Working exercises is one of the main ways of learning mathematics and they are commended to the reader.

There are acknowledgements to be made. I have learned this material in the course of many years and from many sources. It is impossible to list all of the mathematicians who have contributed to my
knowledge on the subject. The material in these notes is well-known,
although the point of view is perhaps fresh, and is mainly drawn from

the books in the bibliography. All of these references have been used. In at least one case, the proof of a theorem is the result of combining three different proofs from these references. This is why we have chosen to make a general acknowledgement rather than give specific references at each point in the text.

Correct method of reckoning, for grasping the meaning of things and knowing everything that is, obscurities ... and all secrets.

Ahmose   16th Century B.C.

This is an English rendering of the title of an Egyptian papyrus. It is taken from the book:

Gay Robins and Charles Shute, "The Rhind Mathematical Papyrus," British Museum Publications, London, 1987.
Quoted by kind permission of the Trustees of the British Museum.

Chapter 1

Calculus on Manifolds

A Lie group is a set which has both a manifold and a group structure. Before we study such objects, we shall review the basic theory of manifolds. As much as possible, we shall work in a coordinate free way.

Let M be a Hausdorff topological space and E a finite dimensional real vector space. An injective map $\varphi : U \to V$, where U is an open set in M and V is one in E, is called a chart. If a point $p \in U$ then φ is called a chart about p. When two charts, $\varphi_\alpha : U_\alpha \to V_\alpha$ and $\varphi_\beta : U_\beta \to V_\beta$ overlap, we get a map on the overlap denoted by $\varphi_{\alpha\beta}$. This is $\varphi_{\alpha\beta} = \varphi_\beta \varphi_\alpha^{-1} : V_{\alpha\beta} \to V_\beta$ where $V_{\alpha\beta}$ is $\varphi_\alpha(U_\alpha \cap U_\beta)$. This overlap map, $\varphi_{\alpha\beta}$, is a map between open subsets of a real vector space and so, we can apply the theory of functions of several real variables to it. Much of the basic work on manifolds is done by working with the overlap functions, $\varphi_{\alpha\beta}$, for all pairs of overlapping charts.

A collection of charts $\{(\varphi)_\alpha, U_\alpha, V_\alpha)\}$ such that $M \subset \cup U_\alpha$ is called an atlas for M.

DEFINITION 1.1   The set M is a manifold modeled on E if it has an atlas.

Notice that the vector space E has been fixed throughout and that $V_\alpha \subset E$ for each α. The dimension of M, $\dim M$, is the dimension of E as a vector space. We say that M is smooth if $\varphi_{\alpha\beta}$ are smooth for all overlap functions $\varphi_{\alpha\beta}$. Similarly M is analytic or C^r if all $\varphi_{\alpha\beta}$ are analytic or C^r. These distinctions are very important in the general theory of manifolds. However, it is a fact that a compact Lie group is automatically analytic and so we shall state all our results only in the case of smooth manifolds. For the remainder of this chapter, all the maps and manifolds considered will be smooth, that is C^∞, unless otherwise stated. This is not a choice of great substance and we could equally well have chosen analytic manifolds instead.

Example 1.2 $M = \mathbf{R}^n$, $E = \mathbf{R}^n$ and the chart $\varphi : M \to \mathbf{R}^n$ the identity map. This is an n-dimensional manifold.

Example 1.3 $M = S^n = \{x \in \mathbf{R}^{n+1} : \sum x_i^2 = 1\}$, $E = \mathbf{R}^n$ and the two charts $U_+ = \{x \in M : x_{n+1} > -1\}$ with $\varphi_+ : U_+ \to \mathbf{R}^n$ by $\varphi_+(x) = \left(\frac{x_1}{1+x_{n+1}}, \ldots, \frac{x_n}{1+x_{n+1}}\right)$ and $U_- = \{x \in M : x_{n+1} < 1\}$ with $\varphi_- : U_- \to \mathbf{R}^n$ by $\varphi_-(x) = \left(\frac{x_1}{1-x_{n+1}}, \ldots, \frac{x_n}{1-x_{n+1}}\right)$. This is an n-dimensional manifold.

Example 1.4 $M = \mathbf{R}$ with the discrete topology and $E = \mathbf{R}^0 = \{p\}$, the one point with charts in $U_x = \{x\}$, $\varphi_x(x) = p$ indexed by $x \in \mathbf{R}$. This is a zero dimensional manifold. Notice that Example 1.2 in the case $n = 1$ and Example 1.4 show that the same underlying set can have two different topologies and two different manifold structures.

From now on we suppose that M is a smooth manifold with a given atlas and that $C^\infty(M)$ is the set of all smooth functions on M. We introduce calculus on M by means of tangent vectors.

DEFINITION 1.5 A map $X : C^\infty(M) \to \mathbf{R}$ is called a tangent vector at $p \in M$ if X satisfies

$$\begin{aligned} &\text{i)} \quad X(f + g) = X(f) + X(g), \\ &\text{ii)} \quad X(fg) = X(f)g(p) + f(p)X(g). \end{aligned}$$

The basic result about tangent vectors is that they form a vector space.

THEOREM 1.6 *The set, $T_p(M)$, of all tangent vectors at p is a finite dimensional vector space with dimension equal to that of M.*

Proof When we define the sum of two tangent vectors and scalar multiplication by a real number it will be clear that $T_p(M)$ is a vector space. The definitions are

$$(X + Y)(f) = X(f) + Y(f) \tag{1.1}$$

and

$$(rX)(f) = r(X(f)), \tag{1.2}$$

where r is a real number and X and Y are tangent vectors at p. To see the result on dimensions we pick a local chart $\varphi : U \to V$. Then taking directional derivatives gives an injection $E \to T_p(M)$:

$$X_v(f) = J\left(f\varphi^{-1}\right)_p(v),$$

where $J\left(f\varphi^{-1}\right)_p$ is the derivative of $f\varphi^{-1}$ at the point p, which in standard coordinates is given by the Jacobian matrix of partial derivatives.
    To see that this is surjective we have to be able to solve

$$J\left(f\varphi^{-1}\right)_p(v) = X(f) \tag{1.3}$$

for v so that (1.3) is true for all f. This is a consequence of Taylor's theorem for functions of several variables.

We now set $T(M) = \cup_p T_p(M)$ the disjoint union over all points of the tangent vectors at each point. It is a fact that $T(M)$ is a vector bundle over M, called the tangent bundle. This means that $T(M)$ can be given a topology so that it is a manifold modeled on $E \times E$ and the projection $\pi : T(M) \to M$ is locally trivial. To make this last statement clear note that $\pi(X) = p$ where $X \in T_p(M)$ then there is an open set $U_p \subseteq M$ so that $\pi^{-1}(U_p) \cong U_p \times E$.

DEFINITION 1.7   A vector field is a section of $T(M)$.

If we denote a vector field by X then

$$X : M \to T(M) \tag{1.4}$$

and X satisfies $\pi(X(p)) = p$ or equivalently $X(p) \in T_p(M)$.

Notice that a vector field, X and a function f give rise to a new function $\tilde{X}(f)$ satisfying $\tilde{X}(f)(p) = X(p)(f)$. Often we use the notation X_p in place of $X(p)$. This gives rise to the following definition of the extension of a vector field.

DEFINITION 1.8   The extension of a vector field X is the map $\tilde{X} : C^\infty(M) \to C^\infty(M)$.

Given two vector fields X and Y we can define their bracket.

DEFINITION 1.9   The bracket of two vector fields is the vector field defined by

$$[X,Y](p)(f) = X(p)(\tilde{Y}(f)) - Y(p)(\tilde{X}(f)).$$

The importance of extensions can be seen from the following results.

Lemma 1.10   *If $\tilde{X} = \tilde{Y}$ then $X = Y$.*

Proof   If $\tilde{X} = \tilde{Y}$ then $X_p(f) = Y_p(f)$ so $X = Y$.

Theorem 1.11   *If X and Y are two vector fields then $[X,Y]^{\tilde{}} = \tilde{X}\tilde{Y} - \tilde{Y}\tilde{X}$.*

Proof   This result follows easily from the definitions. We list the main steps:

$$
\begin{aligned}
[X,Y]^{\tilde{}}(f)(p) &= [X,Y]_p(f) \\
&= X_p(\tilde{Y}f) - Y_p(\tilde{X}f) \\
&= \tilde{X}(\tilde{Y}f)(p) - \tilde{Y}(\tilde{X}f)(p).
\end{aligned}
$$

The result of Theorem 1.11 allows us to list many properties of the bracket operation.

Example 1.12 $[X,X] = 0$.

Example 1.13 $[X,Y] = -[Y,X]$.

That is, the bracket is skew symmetric.

Example 1.14 $[X,[Y,Z]] + [Y,[Z,X]] + [Z,[X,Y]] = 0$.

This is the Jacobi identity.

These examples together with the bilinearity of the bracket say that the set of all vector fields forms a Lie Algebra.

Finally, we discuss the notion of derivatives of f. If $f : M \to N$ is a map between two manifolds, then f is smooth if $\psi_\beta f \varphi_\alpha^{-1}$ is smooth for all charts φ_α of M and ψ_β of N such that the composition is defined. The next question is that of defining the derivative of a smooth map.

Definition 1.15   The derivative of f is $df : T(M) \to T(N)$ with $df(X)(g) = X(g \circ f)$.

To see that this definition makes sense, we note that if $g \in C^\infty(N)$, then $g \circ f \in C^\infty(M)$. Now, if $X \in T(M)$ is a tangent vector at some point $p \in M$, then X acts on $C^\infty(M)$ and, in particular, gives a number $X(g \circ f)$. To define a vector field $df(X) \in T(N)$ at a point $f(p) \in N$, we need to specify the value of $df(X)(g)$ for any function $g \in C^\infty(N)$. This is done by the formula of the definition and is $X(g \circ f)$.

Example 1.16   If $M = N = \mathbb{R}$, then under the natural identification $T_p(\mathbb{R}) \cong \mathbb{R}$ the map $(df)_p : \mathbb{R} \to \mathbb{R}$ is just multiplication by $f'(p)$, where f' is the derivative of elementary calculus.

Exercises - Chapter 1

1.1   Let $M(n)$ be the set of all real $n \times n$ matrices. Show that $M(n)$ is an n^2-dimensional manifold.

1.2   Let X and Y be vector fields on M and f and $g \in C^\infty(M)$. Show $[fX, gY] = f\tilde{X}(g)Y - g\tilde{Y}(f)X + fg[X, Y]$.

1.3   Prove the result of Example 1.12 : $[X, X] = 0$.

1.4   Prove the result of Example 1.13 : $[X, Y] = -[Y, X]$.

1.5   Prove the Jacobi identity: $[X, [Y, Z]] + [Y, [Z, X]] + [Z, [X, Y]] = 0$.

1.6   Let $\varphi : U \to \mathbb{R}^n$ be a chart in M with coordinates $x^1, \ldots, x^n$; that is $\varphi(p) = (x^1, \ldots, x^n)$ for $p \in U$. Let $a^1, \ldots, a^n$ be real numbers and for $f : U \to \mathbb{R}$. Define

$$X(f) = a^1 \frac{\partial}{\partial x^1}(f \circ \varphi^{-1}) + a^2 \frac{\partial}{\partial x^2}(f \circ \varphi^{-1}) + \ldots + a^n \frac{\partial}{\partial x^n}(f \circ \varphi^{-1}).$$

Show that X is a tangent vector.

1.7   With the notation of Exercise 1.6 let $X(p) = \Sigma_i a^i(p) \frac{\partial}{\partial x^i}$ and $Y(p) = \Sigma_i b^i(p) \frac{\partial}{\partial x^i}$, where a^i and b^i are smooth functions U. Then X and Y are two vector fields on U.

Show

$$[X, Y](p) = \sum_{i,j} a^i(p) \frac{\partial}{\partial x^i}\left(b^j \circ \varphi^{-1}\right)(\varphi(p)) \frac{\partial}{\partial x^j}$$

$$- \sum_{i,j} b^i(p) \frac{\partial}{\partial x^i} \left(a^j \circ \varphi^{-1} \right) (\varphi(p)) \frac{\partial}{\partial x^j}.$$

1.8   A curve in M is a function $\gamma : [a, b] \rightarrow M$. If $0 \in [a, b]$ and $\gamma(0) = p$ show that

$$X(f) = \lim_{h \to 0} \frac{f(\gamma(h)) - f(p)}{h}$$

defines a tangent vector to M at p.

1.9   Let $f : \mathbf{R} \rightarrow \mathbf{R}$ be a differentiable function. Show how Definition 1.15 agrees with the usual definition of the derivative of a real valued function of a real variable (that is, give the details in Example 1.16).

1.10   If $M = \mathbf{R}^n$, show how to identify $T_p M$ with $\mathbf{R}^n$ by the formula $X(f) = < grad \, f, v >$ where $X \in T_p M$ is identified with $v \in \mathbf{R}^n$.

Chapter 2

Groups and Lie Groups

A Lie group G is both a manifold and a group. We have discussed manifolds in the previous chapter, and now we turn our attention to group theory.

DEFINITION 2.1   A set G is a group if there is a map $\mu : G \times G \to G$ satisfying

1) $\mu(x, \mu(y, z)) = \mu(\mu(x, y), z)$ for all x, y and z in G;
2) there is an element $1 \in G$ such that $\mu(1, x) = \mu(x, 1) = x$, for all $x \in G$; and
3) for $x \in G$ there exists $y \in G$, such that $\mu(x, y) = \mu(y, x) = 1$.

As a matter of notation we usually suppress μ and write xy in place of $\mu(x, y)$. The equations in the definition now become

$$
\begin{aligned}
&1) \quad x(yz) = (xy)z, &\qquad (2.1)\\
&2) \quad 1x = x1 = x, &\qquad (2.2)\\
&3) \quad xy = yx = 1. &\qquad (2.3)
\end{aligned}
$$

Equation (2.1) is called the associative property. The element 1 of (2.2) is called the identity element of G. The element y in (2.3) is usually written $y = x^{-1}$, the inverse of x. The map μ is called multiplication and the map $\lambda : G \to G$ with $\lambda(x) = x^{-1}$ is the inverse map.

DEFINITION 2.2   A set G is a Lie group if

1) G is an analytic manifold,
2) G is a group,
3)   both μ and λ are analytic maps.

Example 2.3 $G = \mathbb{R}^n$.

Example 2.4 $G = GL(n, \mathbb{R})$, the set of $n \times n$ invertible matrices with real entries.

Example 2.5 $G = GL(n, \mathbb{C})$, the set of $n \times n$ invertible matrices with complex entries.

When the entries of the matrix can be either real or complex but are not specified, we use $GL(n)$ to denote either $GL(n, \mathbb{R})$ or $GL(n, \mathbb{C})$ as appropriate.

We shall show that Example 2.4 really is a Lie group.

Proof of (2.4)   The first step is to give $GL(n, \mathbb{R})$ a manifold structure. If $(a_{ij}) \in GL(n, \mathbb{R})$ let $x_{i+n(j-1)} = a_{ij}$; then the map

$$(a_{ij}) \rightarrow (x_k) \tag{2.4}$$

is an injection $GL(n, \mathbb{R}) \rightarrow \mathbb{R}^{n^2}$. Thus $GL(n, \mathbb{R})$ is identified as a subset of $\mathbb{R}^{n^2}$. If we identify $\mathbb{R}^{n^2}$ with the set of all matrices, invertible or not, in a similar way to map (2.4), then we can introduce the determinant function:

$$\det: \quad \mathbb{R}^{n^2} \rightarrow \mathbb{R}. \tag{2.5}$$

Since det is a polynomial of degree n it is analytic. From elementary linear algebra we know

$$GL(n, \mathbb{R}) \cong \det^{-1}(\mathbb{R}^\star), \tag{2.6}$$

where $\mathbb{R}^\star = \{x \in \mathbb{R} : x \neq 0\}$. Thus, $GL(n, \mathbb{R})$ is an open subset of $\mathbb{R}^{n^2}$, and therefore, it has the structure of an analytic manifold. With this analytic structure, the determinant map is an analytic map.

It remains to check that μ and λ are analytic. Since

$$\mu\left((a_{ij})(b_{ij})\right) = \left(\sum_{k=1}^{n} a_{ik}b_{kj}\right), \tag{2.7}$$

we see that μ is a polynomial map and so is analytic. Cramer's rule expresses λ as a rational function in det and so λ is analytic.

This completes the proof that $GL(n, \mathbb{R})$ is a Lie group.

Example 2.6 $S^1 = \mathrm{R}/\mathrm{Z}$.

We shall prove that S^1 is a Lie group.

Proof of (2.6)   Clearly R is a Lie group and Z is a discrete normal subgroup. Since Z is discrete R/Z is a manifold modeled on R. Since Z is normal R/Z is a group. The analyticity is a local property and so the analyticity of addition on R gives analyticity on R/Z.

COROLLARY 2.7 $T = T^n = \mathrm{R}^n/\mathrm{Z}^n$ *is a Lie group.*

A group G is called commutative or abelian if $xy = yx$ for all pairs of elements x and y of G.

A convenient point to start the study of the theory of Lie groups is at the identity element.

THEOREM 2.8   *Let G_0 be the connected component of the identity element of G. Then G_0 is both a closed and open, normal subgroup of G and is a Lie group.*

Proof   That G_0 is both closed and open is a consequence of elementary point set topology. To see that G_0 is a subgroup we need to show that $G_0^2 \subset G_0$ and $G_0^{-1} \subset G_0$, where $G_0^2 = \{xy : x, y \in G_0\}$ and $G_0^{-1} = \{x^{-1} : x \in G_0\}$. Let $x \in G_0$ then xG_0 is connected and $x \in xG_0$. Thus $xG_0 \cap G_0 \neq 0$ and so $xG_0 \subset G_0$. Since this is true for all $x \in G_0$ it follows that $G_0^2 \subset G_0$. Similarly G_0^{-1} is connected and $G_0^{-1} \cap G_0 \neq 0$, since $1 \in G_0^{-1}$ and $1 \in G_0$, so $G_0^{-1} \subset G_0$. Thus G_0 is a subgroup of G.

The map $y \to xyx^{-1}$ is continuous and so $xG_0x^{-1} \subset G_0$ by the same arguments that show $G_0^2 \subset G_0$. Thus G_0 is a normal subgroup.

That G_0 is a Lie group is an easy consequence of the definitions, which completes the proof.

To move away from the identity element we use left translation. The map
$$L_a : G \to G , \ L_a(g) = ag \qquad (2.8)$$
is left translation by a. The derivative is
$$dL_a : T(G) \to T(G), \qquad (2.9)$$
where $dL_a(T_gG) \subset T_{ag}G$.

DEFINITION 2.9   A vector field X is left invariant if $X \circ L_a = dL_a(X)$.

THEOREM 2.10   *A left invariant vector field is analytic.*

Proof   Let X be a left invariant vector field. Then

$$X(g) = (dL_g)_1 X(1). \tag{2.10}$$

Since L_g is an analytic map of G to itself then the map $g \to (dL_g)_1$ is an analytic map. This proves that X is analytic.

THEOREM 2.11   *The map $X \to X(1)$ defines a one to one correspondence between left invariant vector fields and $T_1(G)$.*

Proof   For $v \in T_1(G)$ define X_v by

$$X_v(g) = (dL_g)_1 (v). \tag{2.11}$$

The map $v \to X_v$ is the inverse map to the map of the theorem.

DEFINITION 2.12   The space $T_1(G)$ is the Lie algebra of G. We usually write $\mathcal{G}$ for $T_1(G)$.

The difficulty with Definition 2.12, is that we have not specified the bracket operation. This is, of course, given by the bracket on vector fields. At the end of Chapter 1, we saw that such a bracket was skew symmetric and satisfied the Jacobi identity, that is $T_1(G)$ satisfied the usual conditions for a Lie algebra. This reduces the problem of showing that the bracket of two left invariant vector fields is again left invariant.

THEOREM 2.13   *Let X and Y be two left invariant fields. Then $[X,Y]$ is left invariant.*

Proof   We use the extension from X and Y to $\tilde{X}$ and $\tilde{Y}$. Left invariance now appears as

$$(\tilde{X} f) \circ L_g = \tilde{X} (f \circ L_g), \tag{2.12}$$

and similarly for $\tilde{Y}$, where in this case $\tilde{X} : C^\infty(G) \to C^\infty(G)$. Left invariance for $[X,Y]$ now follows by the calculation:

$$([X,Y]^\sim f) \circ L_g \;=\; \left(\tilde{X} (\tilde{Y} f) - \tilde{Y} (\tilde{X} f) \right) \circ L_g$$

$$\begin{aligned}
&= \tilde{X}\left(\tilde{Y}f\right)\circ L_g - \tilde{Y}\left(\tilde{X}f\right)\circ L_g \\
&= \tilde{X}\left(\left(\tilde{Y}f\right)\circ L_g\right) - \tilde{Y}\left(\left(\tilde{X}f\right)\circ L_g\right) \\
&= \tilde{X}\left(\tilde{Y}\left(f\circ L_g\right)\right) - \tilde{Y}\left(\tilde{X}\left(f\circ L_g\right)\right) \\
&= [X,Y]^{\sim}\left(f\circ L_g\right).
\end{aligned} \tag{2.13}$$

Thus, the bracket of two left invariant vector fields is, again, left invariant.

Example 2.14   One of the smallest interesting examples of a Lie group is $SU(2)$. This is defined to be the set of 2×2 matrices with complex entries such that the inverse matrix is the complex conjugate of the transpose and with determinant 1. That is

$$SU(2) = \{A : \overline{A}^t = A^{-1} \quad\text{and}\quad \det A = 1\}. \tag{2.14}$$

Let z and w be complex numbers such that $\mid z\mid^2 + \mid w\mid^2 = 1$. Then

$$A = \begin{pmatrix} z & w \\ -\bar{w} & \bar{z} \end{pmatrix} \tag{2.15}$$

is an element of $SU(2)$. Conversely, if $A \in SU(2)$ then there exist z and w so that A has the form given in equation (2.15) . Let $z = x + iy$ and $w = u + iv$ so that $x^2 + y^2 + u^2 + v^2 = 1$. Then we can define charts on $SU(2)$ as follows:

$$\begin{aligned}
U_1 &= \{A : x > 0\}, \; U_2 = \{A : x < 0\}, \\
U_3 &= \{A : y > 0\}, \; U_4 = \{A : y < 0\}, \\
U_5 &= \{A : u > 0\}, \; U_6 = \{A : u < 0\}, \\
U_7 &= \{A : v > 0\}, \; U_8 = \{A : v < 0\}.
\end{aligned} \tag{2.16}$$

The corresponding maps are given by

$$\begin{aligned}
\varphi_1(A) &= (y, u, v), \; \varphi_2(A) = (y, u, v), \\
\varphi_3(A) &= (x, u, v), \; \varphi_4(A) = (x, u, v), \\
\varphi_5(A) &= (x, y, v), \; \varphi_6(A) = (x, y, v), \\
\varphi_7(A) &= (x, y, u), \; \varphi_8(A) = (x, y, u).
\end{aligned} \tag{2.17}$$

It is an easy check to see that the overlap maps are analytic. Let $\varphi : SU(2) \to \mathbb{R}^4$ by

$$\varphi(A) = (x, y, u, v). \tag{2.18}$$

Then if A_1 and A_2 are two matrices given by (z_1, w_1) and (z_2, w_2) so $\varphi(A_i) = (x_i, y_i, u_i, v_i)$ then

$$\varphi(A_1 A_2) = (x_3, y_3, u_3, v_3) \tag{2.19}$$

with

$$
\begin{aligned}
x_3 &= x_1 x_2 - y_1 y_2 - u_1 u_2 - v_1 v_2, \\
y_3 &= x_1 y_2 + x_2 y_1 + u_1 v_2 - u_2 v_1, \\
u_3 &= x_1 u_2 + x_2 u_1 - y_1 v_2 + y_2 v_1, \\
v_3 &= x_1 v_2 + x_2 v_1 + y_1 u_2 - y_2 u_1.
\end{aligned} \tag{2.20}
$$

Since the maps φ_i are the composition $P_i \varphi \mid U_i$, where P_i is the appropriate projection of $\mathbf{R}^4$ onto various 3-dimensional subspaces, it is clear that matrix multiplication is an analytic map. It is even easier to see that the inverse map is analytic:

$$\varphi(A^{-1}) = (x, -y, -u, -v). \tag{2.21}$$

Thus $SU(2)$ is a 3-dimensional Lie group. Notice that the map φ defined by equation (2.18) is a diffeomorphism from $SU(2)$ onto S^3— the set of unit vectors in $\mathbf{R}^4$. This space S^3 is the three-dimensional sphere.

Example 2.15   There are a number of other classical groups of which the most commonly encountered ones are the following

   a)   The unitary groups $U(n)$.

These are the $n \times n$ matrices with complex entries such that the inverse is the conjugate transpose. That is,

$$U(n) = \{A : \overline{A}^t = A^{-1}\}. \tag{2.22}$$

   b)   The special unitary groups $SU(n)$.
These are subgroups of $U(n)$ with determinant one. That is,

$$SU(n) = \{A : \overline{A}^t = A^{-1} \text{ and } \det A = 1\}. \tag{2.23}$$

   c)   The orthogonal groups $O(n)$.
These are the $n \times n$ matrices with real entries such that the inverse is the transpose:

$$O(n) = \{A : A^t = A^{-1}\}. \tag{2.24}$$

d) The special orthogonal groups $SO(n)$.
These are the subgroups of $O(n)$ with determinant one:

$$SO(n) = \{A : A^t = A \text{ and } \det A = 1\}. \tag{2.25}$$

e) The symplectic group $Sp(n)$.

Let H denote the quaternions. Then $H \cong R^4$ and an element of H is written as $x + iy + jz + kw$. Conjugation of a quaternion is given by

$$\overline{x + iy + ju + kv} = x - iy - ju - kv. \tag{2.26}$$

The group $Sp(n)$ consists of $n \times n$ matrices with quaternion entries such that the inverse is the conjugate transpose. That is,

$$Sp(n) = \{A : \overline{A}^t = A^{-1}\}. \tag{2.27}$$

Notice that $Sp(1)$ is the set of unit quaternions so that $Sp(1)$ is diffeomorphic to S^3. Combining this observation with the corresponding one about $SU(2)$, which was made after equation (2.21), we see that $SU(2)$ and $Sp(1)$ are diffeomorphic as manifolds. Now, if we multiply two quaternions, we find:

$$\begin{aligned}(x_1 + iy_1 + ju_1 + kv_1)&(x_2 + iy_2 + ju_2 + kv_2) \\ = \ x_3 + iy_3 &+ ju_3 + kv_3\end{aligned} \tag{2.28}$$

where x_3, y_3, u_3, and v_3 are given by equations (2.20). Thus φ is not just a diffeomorphism between $SU(2)$ and $Sp(1)$ it is also a group isomorphism.

Exercises - Chapter 2

2.1 Show that $SO(3)$ is a Lie group.

2.2 Let c^* be the non-zero complex numbers and define $\varphi : c^* \rightarrow GL(2, R)$ by $\varphi\left(re^{i\theta}\right) = r \begin{pmatrix} \cos\theta & -\sin\theta \\ \sin\theta & \cos\theta \end{pmatrix}$. Show $\varphi(zw) = \varphi(z)\varphi(w)$ and $\det \varphi(z) = |z|^2$.

2.3 Let $A \in U(n)$. Then show $|\det A| = 1$.

2.4 Let $A \in O(n)$. Then show $\det A = 1$ or $\det A = -1$.

2.5   Let w and z be complex numbers such that $w\overline{w} + z\overline{z} = 1$ where $\overline{w}$ is the complex conjugate of w. Let

$$A = \begin{pmatrix} z & w \\ -e^{i\theta}\,\overline{w} & e^{i\theta}\,\overline{z} \end{pmatrix}.$$ Then show $A \in U(2)$ and $\det A = e^{i\theta}$.

2.6   (Converse to Exercise 2.5) Let $A \in U(2)$. Then, show there are w and $z \in$ C such that $w\overline{w} + z\overline{z} = 1$ and $\theta \in$ R, so

$$A = \begin{pmatrix} z & w \\ -e^{i\theta}\,\overline{w} & e^{i\theta}\,\overline{z} \end{pmatrix}.$$

2.7   Use the results of Exercises 2.5 and 2.6 to show that $U(2)$ is a four-dimensional Lie group.

2.8   Let $A \in SO(3)$ show that A has at least one eigenvalue which is $+1$. Hence, deduce that A is similar to a matrix of the form

$$\begin{pmatrix} \cos\theta & -\sin\theta & 0 \\ \sin\theta & \cos\theta & 0 \\ 0 & 0 & 1 \end{pmatrix}$$

for some θ. (A is similar to B if there is a matrix P with $A = PBP^{-1}$.)

2.9   Let $A \in O(3)$ but $A \notin SO(3)$. Show that A has at least one eigenvalue which is -1. Hence deduce that A is similar to a matrix of the form

$$\begin{pmatrix} \cos\theta & -\sin\theta & 0 \\ \sin\theta & \cos\theta & 0 \\ 0 & 0 & -1 \end{pmatrix}$$

for some θ.

2.10   Let $A \in SU(2)$ and λ be an eigenvalue for A. Show $\lambda = e^{i\theta}$ for some θ and deduce that A is similar to a matrix of the form

$$\begin{pmatrix} e^{i\theta} & 0 \\ 0 & e^{-i\theta} \end{pmatrix}.$$

Chapter 3

One-Parameter Subgroups and the Exponential Map

In the previous chapter, we saw that the Lie algebra of G could be realized as the tangent space at the identity or as the left invariant vector fields. Now, we shall see a third description of the Lie algebra as the set of one-parameter subgroups .

DEFINITION 3.1   An analytic homomorphism $f : \mathbf{R} \to G$ is a one-parameter subgroup of G.

By homomorphism we mean that f satisfies

$$f(x + y) = f(x)f(y). \tag{3.1}$$

THEOREM 3.2   *The map $f \to (df)_0(1)$ is a one to one correspondence between one-parameter subgroups of G and $T_1(G)$.*

Proof   Let $v \in T_1(G)$ and define the left invariant vector field X_v as before:

$$X_v(g) = (dL_g)_1(v). \tag{3.2}$$

By a classical result on differential equations there is an open interval $J \subset \mathbf{R}$ with $0 \in J$ and a function $f : J \to G$ such that

$$(df)_t(1) = X_v(f(t)). \tag{3.3}$$

Let $J_1 \subset J$ such that $J_1 + J_1 \subset J$ and fix $s \in J_1$. Then for $t \in J_1$ the following two functions

$$t \to f(s)f(t) \tag{3.4}$$

and

$$t \to f(s + t) \tag{3.5}$$

15

both satisfy equation (3.3). Thus, by the uniqueness of the solution to the differential equation we have

$$f(s)f(t) = f(s + t) \qquad (3.6)$$

for s and t both in J_1. Now extend f to all of $\mathbb{R}$ by the formula

$$f(t) = f(t/N)^N \qquad (3.7)$$

for suitably large N. This function is a one parameter subgroup and we denote f by f_v. The map $v \to f_v$ is the inverse of the map $f \to (df)_0(1)$, which completes the proof.

We keep the notation f_v for the solution of equation (3.3) and use it to define the exponential map. First, we observe the relationship between f_{sv} and f_v. Let g_{sv} be the one-parameter subgroup $g_{sv}(t) = f_v(st)$. Then by the chain rule

$$(dg_{sv})_t = s\,(df_v)_{st}\,. \qquad (3.8)$$

Hence g_{sv} is the one-parameter subgroup associated to sv. Thus we have

$$f_{sv}(t) = f_v(st). \qquad (3.9)$$

DEFINITION 3.3   The exponential map is $\exp : \mathcal{G} \to G$ where $\exp(v) = f_v(1)$.

THEOREM 3.4   *There are open neighborhoods $V \subset \mathcal{G}$ of 0 and $U \subset G$ of 1 such that $\exp | V$ is an analytic homeomorphism from V to U.*

Proof   We have defined $\exp$ by solving a differential equation. The existence of a solution of the differential equation was guaranteed by a classical theorem. Analyticity of $\exp$ is given by the same theorem.

The existence of U and V and the fact that $\exp$ is a homeomorphism are trivial consequences of the inverse function theorem once we show that $d(\exp)_0$ is non-singular. To prove this we need the formula

$$v(f) = \lim_{h \to 0} \frac{f\left(f_v(h)\right) - f(1)}{h}, \qquad (3.10)$$

which we prove as a lemma below. If we identify $T_0(\mathcal{G})$ with $\mathcal{G}$ we find that $d(\exp)_0 : \mathcal{G} \to \mathcal{G}$.

Now we calculate

$$d\left(\exp_0\right)vf = \lim_{t\to 0}\frac{f(\exp(tv)) - f(1)}{t} \qquad (3.11)$$

$$= \lim_{t\to 0}\frac{f(f_{tv}(1)) - f(1)}{t} \qquad (3.12)$$

$$= \lim_{t\to 0}\frac{f(f_v(t)) - f(1)}{t} \qquad (3.13)$$

$$= v(f). \qquad (3.14)$$

Here (3.11) is an immediate consequence of (3.10). The definition of exp gives (3.12); and (3.13) follows from (3.9). Finally, (3.14) is just (3.10) used again, and the identification of one-parameter subgroups with $\mathcal{G}$. Hence, we have $d(\exp_0)v = v$ and so $d(\exp_0)$ is the identity map which is clearly nonsingular. The proof of this theorem is now completed when we prove the next lemma.

LEMMA 3.5 *Let X_v be the left invariant vector field and f_v be the one-parameter subgroup associated with v . Then for $f : G \to$ R*

$$X_v\left(f_v(t)\right)(f) = \lim_{h\to 0}\frac{f\left(f_v(t+h)\right) - f\left(f_v(t)\right)}{h}.$$

Proof Define a vector field Y on $f_v(R)$ by

$$Y\left(f_v(t)\right)f = \lim_{h\to 0}\frac{f\left(f_v(t+h)\right) - f\left(f_v(t)\right)}{h}. \qquad (3.15)$$

That Y is a vector field is equivalent to the definition of a derivative in elementary calculus; see also Exercise 1.8. It is clear from (3.15) that

$$dL_{f_v(t)}Y(1)f = Y\left(f_v(t)\right)f. \qquad (3.16)$$

Thus Y is left invariant along $f_v(\text{R})$. Now

$$(df_v)\,(1)f = 1\,(f \circ f_v) \qquad (3.17)$$

by definition where 1 is the unit vector in $T(\text{R})$. Thus, by Example 1.16 we have

$$(df_v)_t\,(1)f = \lim_{h\to 0}\frac{f\left(f_v(t+h)\right) - f\left(f_v(t)\right)}{h}. \qquad (3.18)$$

Combining (3.15) and (3.18) gives that

$$(df_v)_t(1) = Y(f_v(t)),$$
(3.19)

or that f_v is the one-parameter subgroup associated with $Y(1)$. By the uniqueness of solutions of (3.3) this implies that $Y(1) = v$ which completes the proof of the lemma.

A subgroup H of G is called a small subgroup if $H \subset U$ for every open set U in G which contains the identity element.

COROLLARY 3.6   *The Lie group G has no small subgroups.*

Proof   Let U and V be as in Theorem 3.4 and let H be a subgroup of G such that $H \subset U$ and let $g \in H$ with $g \neq 1$. This implies $g \in U$. Then there is $v \in V$ such that $\exp v = g$. Since $\mathcal{G}$ is a vector space, pick an open neighborhood W of 0 such that $W \subset V$ and $v \notin W$. Then $\exp W$ is an open neighborhood of 1 in G such that $g \notin \exp W$. Thus G has no non-trivial subgroups contained in all open neighborhoods of the identity.

The structure of Lie groups imposes strong properties on maps between groups. We see this in the following theorem.

THEOREM 3.7   *Let $f : G \to H$ be a continuous homomorphism. Then f is analytic.*

Proof   It is only necessary to prove that f is analytic in a neighborhood of 1 in G.

We start with continuous one-parameter subgroups, that is, the case $G = \mathbf{R}$. Let V be an open neighborhood of 0 in $T_1(H)$ such that $\exp | V$ is a diffeomorphism onto $\exp(V)$ in H. Let $y \in \exp\left(\frac{1}{2}V\right)$. Then there is Y such that

$$y = \exp Y.$$
(3.20)

If $z = \exp\left(\frac{1}{2}Y\right)$ then

$$z^2 = y$$
(3.21)

and $z \in \exp\left(\frac{1}{2}V\right)$. Since $f(0) = 1$ and f is continuous, there is $\epsilon > 0$ such that

$$f(t) \in \exp\left(\frac{1}{2}V\right) \quad \text{for} \quad |t| \leq \epsilon.$$
(3.22)

Suppose $f(\epsilon) = \exp(X)$ then $f(\epsilon/2) = \exp(X/2)$ and by induction

$$f\left(\epsilon/2^n\right) = \exp\left(X/2^n\right). \tag{3.23}$$

Thus, for all integers m and n we have

$$f\left(m\epsilon/2^n\right) = \exp\left(mX/2^n\right). \tag{3.24}$$

Since f is continuous, we have that for r real

$$f(r\epsilon) = \exp(rX) \quad \text{for} \quad |r| \leq 1. \tag{3.25}$$

Thus f is analytic.

We now proceed to the general case. Let $X_1, \ldots, X_n$ be a basis for $\mathcal{G}$. Then by the first part the maps

$$t \rightarrow f\left(\exp\left(tX_j\right)\right) \tag{3.26}$$

are analytic one-parameter subgroups. So we pick $Y_1, \ldots, Y_n$ in $T_1(H)$ such that

$$f\left(\exp\left(tX_j\right)\right) = \exp tY_j \tag{3.27}$$

Define $F : \mathbf{R}^n \rightarrow G$ by the equation

$$F\left(t_1, \ldots, t_n\right) = \exp\left(t_1 X_1\right) \ldots \exp\left(t_n X_n\right). \tag{3.28}$$

Then there is a neighborhood $U \subset \mathbf{R}^n$ such that $F \mid U$ is an analytic diffeomorphism. By equation (3.27) and the fact that f is a homomorphism, we see that $f \circ F$ is analytic. Now the equation

$$f = (f \circ F) \circ F^{-1} \tag{3.29}$$

shows that f is analytic.

We now need to make some remarks on the proof of this theorem. In Definition 3.1, we needed to include analyticity of one-parameter subgroups, so that we could differentiate them, for example, in Theorem 3.2 and its proof. Then, all one-parameter subgroups are described by the equation

$$f_v(t) = \exp(tv), \tag{3.30}$$

which is clear for analytic homomorphisms when we obtain equation (3.9). That any continuous homomorphism also satisfies (3.30) requires more work and is not established until equation (3.25).

The coordinates on G given by (3.28) are very similar to normal coordinates. Had the map F been

$$(t_1, \ldots, t_n) \rightarrow \exp(tX_1 + \ldots + tX_n) \tag{3.31}$$

then we would have had normal coordinates. The difference between (3.28) and (3.31) is the subject of the Campbell-Baker-Hausdorff formula of the next chapter.

Example 3.8   Let A be an $n \times n$ matrix and define $f : \mathrm{R} \rightarrow GL(n)$ by

$$f(t) = I + tA + \frac{t^2 A^2}{2!} + \frac{t^3 A^3}{3!} + \ldots \tag{3.32}$$

Then, since $f(s + t) = f(s)f(t)$ and $(df)_0(1) = A$, we see that f is a one-parameter subgroup of $GL(n)$. Thus, for a matrix group the exponential map is

$$\exp(A) = I + A + \frac{A^2}{2!} + \frac{A^3}{3!} + \ldots, \tag{3.33}$$

and the Lie algebra is a subspace of the vector space of all matrices. Notice that all of these remarks hold for matrices with entries that are real, complex or quaternion. The Lie bracket is then given by

$$[A, B] = AB - BA. \tag{3.34}$$

LEMMA 3.9 $\det(\exp A) = \exp(tr A)$.

Proof   First we prove this in the case when A is an upper triangular matrix with diagonal entries $\lambda_1, \ldots, \lambda_n$. Then A^k is also upper triangular, but this time with diagonal entries $\lambda_1^k, \lambda_2^k, \ldots, \lambda_n^k$. Thus, $\exp A$ is upper triangular with diagonal entries $e^{\lambda_1}, e^{\lambda_2}, \ldots, e^{\lambda_n}$, and we have

$$\begin{aligned}
\det \exp A &= e^{\lambda_1} e^{\lambda_2} \ldots e^{\lambda_n} \\
&= e^{\lambda_1 + \lambda_2 + \ldots + \lambda_n}, \tag{3.35}
\end{aligned}$$

which is the result of the lemma in this special case.

The next case is when $A = PTP^{-1}$ and T is upper triangular. We observe that

$$\exp A = P(\exp T)P^{-1} \tag{3.36}$$

and so $\det \exp A = \det \exp T$, and since $\mathrm{tr}\, A = \mathrm{tr}\, T$ the result again follows.

Finally, we observe that by the Jordan canonical form (extending real coefficients to complex ones, if necessary) any matrix A has the form PTP^{-1} for an upper triangular matrix T.

Example 3.10 The Lie algebra of $SU(2)$ is denoted by $\mathcal{SU}(2)$. This is the set of 2×2 matrices with trace zero, which are skew hermitian. Thus, if $A \in \mathcal{SU}(2)$, then

$$A = \begin{pmatrix} ix & y + iz \\ -y + iz & -ix \end{pmatrix}. \tag{3.37}$$

Let $\varphi : \mathcal{SU}(2) \to \mathbf{R}^3$ by $\varphi(A) = (x, y, z)$. Then, the Lie bracket of $\mathcal{SU}(2)$ is given by

$$[A_1, A_2] = \begin{pmatrix} ix_3 & y_3 + iz_3 \\ -y_3 + iz_3 & -ix_3 \end{pmatrix}, \tag{3.38}$$

where

$$\begin{aligned} x_3 &= 2\,(y_1 z_2 - y_2 z_1)\,, \\ y_3 &= 2\,(z_1 x_2 - z_2 x_1)\,, \\ z_3 &= 2\,(x_1 y_2 - x_2 y_1)\,. \end{aligned} \tag{3.39}$$

Thus, if $\times$ denotes the usual cross product $\times : \mathbf{R}^3 \times \mathbf{R}^3 \to \mathbf{R}^3$ we have

$$\varphi([A_1, A_2]) = 2\varphi(A_1) \times \varphi(A_2). \tag{3.40}$$

Example 3.11 We shall now write down the Lie algebras of a number of classical Lie groups. From this description we shall be able to find the dimension of the group very easily, which is the same as the dimension of the Lie algebra.

a) The unitary groups $U(n)$ have Lie algebra $\mathcal{U}(n)$:

$$\mathcal{U}(n) = \{A : \bar{A}^t = -A\}, \tag{3.41}$$

that is, $\mathcal{U}(n)$ consists of skew hermitian complex matrices. Counting the entries above the diagonal, we find $\frac{1}{2}n(n-1)$ such complex numbers. The diagonal entries are all pure imaginary and the entries below the diagonal are determined by those above. Thus, $\dim U(n) = n + 2\frac{1}{2}n(n-1)$ or

$$\dim U(n) = n^2. \tag{3.42}$$

b)   The special unitary groups $SU(n)$ have Lie algebras $\mathcal{SU}(n)$ which consist of skew hermitian complex matrices with trace zero:

$$\mathcal{SU}(n) = \{A : \bar{A}^t = -A \quad \text{and} \quad tr A = 0\}. \tag{3.43}$$

Thus, counting dimensions as before gives

$$\dim SU(n) = n^2 - 1. \tag{3.44}$$

c)   The orthogonal groups $O(n)$ have Lie algebras $\mathcal{O}(n)$ which consist of skew symmetric matrices with real entries. This is

$$\mathcal{O}(n) = \{A : A^t = -A\}. \tag{3.45}$$

Thus, we have

$$\dim O(n) = \frac{1}{2}n(n-1). \tag{3.46}$$

d)   The special orthogonal groups $SO(n)$ have the same Lie algebra as the orthogonal groups. This is because $SO(n)$ is the connected component of the identity in $O(n)$. For completeness, we record:

$$\dim SO(n) = \frac{1}{2}n(n-1). \tag{3.47}$$

e)   The symplectic groups $Sp(n)$ have Lie algebra $\mathcal{Sp}(n)$ which consists of all skew symplectic matrices:

$$\mathcal{Sp}(n) = \{A : \bar{A}^t = -A\}. \tag{3.48}$$

Counting dimensions in a similar way to part a gives

$$\dim \mathcal{Sp}(n) = n(2n+1). \tag{3.49}$$

Exercises - Chapter 3

3.1   For the group $SU(2)$ calculate $\exp\begin{pmatrix} ix & 0 \\ 0 & -ix \end{pmatrix}$.

3.2   For $SU(2)$ calculate $\exp\begin{pmatrix} 0 & z \\ -\bar{z} & 0 \end{pmatrix}$, where $z \in \mathbb{c}$, using the series (3.33).

3.3   Show that for any real r and θ the map $f(t)$
$$= \begin{pmatrix} \cos(tr) & e^{i\theta}\sin(tr) \\ -e^{-i\theta}\sin(tr) & \cos(tr) \end{pmatrix}$$ is a one-parameter subgroup of $SU(2)$.

3.4   Use Exercise 3.3 to check the result of Exercise 3.2.

3.5   Find P and P^{-1} so that $\begin{pmatrix} ix & z \\ -\bar{z} & -ix \end{pmatrix} = P \begin{pmatrix} i\lambda & 0 \\ 0 & -i\lambda \end{pmatrix} P^{-1}$
where $\lambda = \sqrt{x^2 + z\bar{z}}$.

3.6   Use the result of Exercise 3.5 to show

$$\exp \begin{pmatrix} ix & z \\ -\bar{z} & -ix \end{pmatrix} = \begin{pmatrix} \cos \lambda + i\frac{x}{\lambda} \sin \lambda & \frac{z}{\lambda} \sin \lambda \\ -\frac{\bar{z}}{\lambda} \sin \lambda & \cos \lambda - \frac{ix}{\lambda} \sin \lambda \end{pmatrix}.$$

Hence, observe $\exp \begin{pmatrix} ix & 0 \\ 0 & -ix \end{pmatrix} \exp \begin{pmatrix} 0 & z \\ -\bar{z} & 0 \end{pmatrix} \neq \exp \begin{pmatrix} ix & z \\ -\bar{z} & -ix \end{pmatrix}.$

(This last observation is the reason for studying the Campbell-Baker-Hausdorff formula of the next chapter.)

3.7   Show that if $f : \mathbf{R} \to G$ is a one-parameter subgroup and $g\epsilon G$, then $f^g(t) = gf(t)g^{-1}$ is also a one-parameter subgroup.

3.8   Let H be a subgroup of G and $f : \mathbf{R} \to G$ a one-parameter subgroup. Suppose there is $\epsilon > 0$ so that $f(-\epsilon, \epsilon) \subset H$. Then show $f(\mathbf{R}) \subset H$. Deduce that if I is any open interval of $\mathbf{R}$ and $f(I) \subset H$ then $f(\mathbf{R}) \subset H$.

3.9   Let g be an element of the group $SU(n)$ and A an element of the Lie algebra $\mathcal{SU}(n)$. Show gAg^{-1} is an element of the Lie algebra of $\mathcal{SU}(n)$ where we have used the usual matrix multiplication.

3.10   Let H be a subgroup of G. Then the Lie algebra $\mathcal{H}$ of H is a subalgebra of the Lie algebra $\mathcal{G}$ of G. Let $\exp_H : \mathcal{H} \to H$ and $\exp_G : \mathcal{G} \to G$ be the exponential maps. Then show $\exp_H = \exp_G \,|\, \mathcal{H}$.

Chapter 4

The Campbell-Baker-Hausdorff Formula

The aim of this chapter is to relate the product $\exp X \exp Y$ to the exponential of the sum $X + Y$. This relation is given by the following theorem, the Campbell-Baker- Hausdorff theorem.

THEOREM 4.1 *Let X and $Y \in \mathcal{G}$. Then there exist $\epsilon > 0$ a real number, $I = (-\epsilon, \epsilon)$ and $Z : I \to \mathcal{G}$ such that for $t \in I$*

$$\exp(tX)\exp(tY) = \exp(Z(t)),$$

and furthermore $Z(t) = \sum_{m=1}^{\infty} t^m Z_m(X,Y)$ where $Z_{m+1}(X,Y)$ is an m-order Lie bracket of X and Y with

$$
\begin{aligned}
Z_1(X,Y) &= X + Y, \\
Z_2(X,Y) &= \frac{1}{2}[X,Y],
\end{aligned}
$$

and

$$Z_3(X,Y) = \frac{1}{12}[[X,Y],Y] - \frac{1}{12}[[X,Y],X].$$

Proof   Define $Z(t) = \exp^{-1}(\exp tX \exp tY)$ and then by Theorem 3.4 we know that for t sufficiently small Z is an analytic function of t. Hence $Z(t) = \sum t^m Z_m(X,Y)$. To identify the terms of this series, we pick an analytic test function f. Then by (3.10) we see that

$$\tilde{X}(\ldots \tilde{X}(f)\ldots) = \tilde{X}^n f = \frac{d^n}{dt^n}f(\exp tX)\,|_{t=0}, \qquad (4.1)$$

where $\tilde{X}$ is the extension of X. Thus

$$\left(\tilde{X}^n \tilde{Y}^m f\right) = \frac{d^n}{dt^n}\frac{d^m}{ds^m} f(\exp tX \exp sY)\,|_{\substack{s=0 \\ t=0}} \tag{4.2}$$

so the Taylor's series of f is

$$f(\exp tX \exp sY) = \sum_{m,n\geq 0} \frac{t^n}{n!}\frac{s^m}{m!}\left(\tilde{X}^n \tilde{Y}^m f\right)(1). \tag{4.3}$$

On the other hand

$$f\left(\exp\left(\sum t^m Z_m(X,Y)\right)\right)$$
$$= f\left(\exp\left(tZ_1 + t^2 Z_2 + t^3 Z_3\right)\right) + O\left(t^4\right) \tag{4.4}$$
$$= \sum \frac{1}{n!}\left(t\tilde{Z}_1 + t^2\tilde{Z}_2 + t^3\tilde{Z}_3\right)^n f(1) + O\left(t^4\right). \tag{4.5}$$

Putting $s = t$ in (4.3) and equating coefficients of t, t^2 and t^3 give the result of the theorem.

We shall now give some applications of this formula. In practice, one rarely needs the expression for Z_3. This will be clear from the following examples.

THEOREM 4.2   *Let H be a closed subgroup of a Lie group G. Then H is a Lie group.*

Proof   Let $\mathcal{H} = \{X \in \mathcal{G} : \exp tX \in H \text{ for all } t \in \mathbf{R}\}$. Then we shall show that $\mathcal{H}$ is a Lie subalgebra of $\mathcal{G}$. Clearly $X \in \mathcal{H}$ if and only if $tX \in \mathcal{H}$ for all $t \in \mathbf{R}$. Suppose X and $Y \in \mathcal{H}$. Then

$$\exp\left(\frac{t}{n}X\right)\exp\left(\frac{t}{n}Y\right) = \exp\left(\frac{t}{n}(X+Y) + O\left(n^{-2}\right)\right). \tag{4.6}$$

So

$$\left(\exp\left(\frac{tX}{n}\right)\exp\left(\frac{tY}{n}\right)\right)^n = \exp\left(t(X+Y) + O\left(n^{-1}\right)\right). \tag{4.7}$$

Since H is closed

$$\lim_{n\to\infty}\left(\exp\left(\frac{tX}{n}\right)\exp\left(\frac{tY}{n}\right)\right)^n \in H. \tag{4.8}$$

Thus

$$\exp(t(X+Y)) \in H. \tag{4.9}$$

That is $X + Y \in \mathcal{H}$. Notice that (4.9) is true for any t since we can choose n large enough to make (4.6) valid.
Similarly, we use the formula

$$
\begin{aligned}
\exp\left(t^2\,[X,Y]\right) \\
= \lim_{n\to\infty}\left(\exp\left(\frac{-tX}{n}\right)\exp\left(\frac{-tY}{n}\right)\exp\left(\frac{tX}{n}\right)\exp\left(\frac{tY}{n}\right)\right)^n
\end{aligned}
\tag{4.10}
$$

to show that $[X,Y] \in \mathcal{H}$. Now, we have shown that $\mathcal{H}$ is a Lie subalgebra of $\mathcal{G}$. Pick a subspace $L \subset \mathcal{G}$ so that

$$
\mathcal{G} = L \oplus \mathcal{H}.
\tag{4.11}
$$

In the text that follows, we show that there is a neighborhood of W of 0 in L such that $X \in W$ implies $\exp X \notin H$. Suppose this is not true. Then, there is a sequence $(X_n) \subset L$ such that $X_n \to 0$ and $\exp X_n \in H$. Pick an Euclidean norm on L; then $(X_n/ \parallel X_n \parallel)$ has a limit point, say X, in L and $\parallel X \parallel = 1$ so $X \neq 0$. Given t we find integers n_m so that

$$
\lim_{m\to\infty} n_m \parallel X_m \parallel = t.
\tag{4.12}
$$

Then

$$
\begin{aligned}
\exp tX &= \lim \exp t\frac{X_m}{\parallel X_m \parallel}
&\tag{4.13}\\
&= \lim \exp n_m X_m
&\tag{4.14}\\
&= \lim \left(\exp X_m\right)^{n_m}.
&\tag{4.15}
\end{aligned}
$$

But H is closed, so $\exp tX \in H$ and so $X \in \mathcal{H}$ which is impossible. Thus, the neighborhood W exists.

Pick a neighborhood V in $\mathcal{G}$ so that $\exp \mid V$ is a diffeomorphism. Let $U = V \cap \mathcal{H}$ and $U' = V \cap L \cap W$. Then, the map

$$
\varphi : U \oplus U' \to G
\tag{4.16}
$$

which is given by

$$
X_1 \oplus X_2 \to \exp X_1 \exp X_2
\tag{4.17}
$$

is a diffeomorphism onto its image. If $\varphi\left(X_1 \oplus X_2\right) \in H$ then $\exp X_2 \in H$ so $X_2 = 0$. Thus φ maps $U \oplus \{0\}$ diffeomorphically onto a neighborhood of 1 in H. Thus H is a submanifold of G and so H is a Lie subgroup of G.

Remark 4.3   It turns out that $\mathcal{H}$ is the Lie algebra of H.

Example 4.4   In Example 2.4, we saw that $GL(n, \mathsf{R})$ is a Lie group. Since O_n and $SO(n)$ are both closed subgroups of $GL(n, \mathsf{R})$ we see immediately that they are Lie groups. While this fact has been stated before and used, we now have a proof of it. To see that $O(n)$ and $SO(n)$ are closed subgroups, consider the following function $f : GL(n, \mathsf{R}) \to M(n)$ the set of $n \times n$ matrices

$$f(A) = A^{-1} - A^t. \tag{4.18}$$

Then $O(n) = f^{-1}\{0\}$ so $O(n)$ is closed being the inverse image of a closed set $\{O\}$, under a continuous map f. Similarly, for $SO(n)$ consider det : $O(n) \to \{1, -1\}$; then $SO(n) = \det^{-1}\{1\}$. Thus $SO(n)$ is a closed subgroup of $O(n)$ and hence, of $GL(n, \mathsf{R})$.

Example 4.5   In an analogous way, we see that $U(n)$ is a closed subgroup of $GL(n, \mathsf{c})$ and $SU(n)$ a closed subgroup of $U(n)$. Similarly $Sp(n)$ is a closed subgroup of $GL(n, \mathsf{H})$.

We now give the correspondence between subgroups of G and subalgebras of $\mathcal{G}$. First, we describe the notion of an immersion.

A map $f : M \to N$ between manifolds with $\dim M \le \dim N$ is an immersion if, at each point $p \in M$, df_p has rank $\dim M$. If f is an immersion, the image $f(M)$ is an immersed submanifold. Note that an immersed submanifold does not need to be a submanifold; see Example 4.7 below.

THEOREM 4.6   *There is a one to one correspondence between immersed subgroups of G and subalgebras of $\mathcal{G}$.*

Proof   If $j : H \to G$ is an immersion and a group homomorphism, then dj_1 is an injective Lie algebra homomorphism. Thus, the image of dj_1 is the required subalgebra.

Conversely, suppose $\mathcal{H}$ is a subalgebra of $\mathcal{G}$. Let H be the subgroup generated by $\exp \mathcal{H}$. To make this into a manifold, we let $X_1, \ldots, X_n$ be a basis for $\mathcal{G}$ such that $X_1, \ldots, X_r$ is a basis for $\mathcal{H}$. Then, locally we define the chart by $\varphi(\exp(x_1 X_1, + \ldots x_r X_r)) = (x_1, \ldots, x_r)$. The inclusion map $j : H \to G$ is an immersion.

Example 4.7   Let $G = T^2$, the two-dimensional torus, and $H = \{e^{it}, e^{i\sqrt{2}\,t}\}$. So H is an irrational line on the torus. Then H is an immersed subgroup, but is not a submanifold of G.

DEFINITION 4.8   A Lie algebra $\mathcal{G}$ is commutative, if $[X, Y] = 0$ for all X and Y in $\mathcal{G}$.

THEOREM 4.9   *A connected Lie group G is commutative if and only if $\mathcal{G}$ is.*

Proof   Pick neighborhoods U of O in $\mathcal{G}$ and V of 1 in G so that $\exp : U \to V$ is a diffeomorphism. Let $W \subset V$ be open, so that $W \cdot W \subset V$. Then, if $x, y \in W$ there are $X, Y \in U$ such that $x = \exp X$ and $y = \exp Y$. Now, suppose the Lie algebra $\mathcal{G}$ is commutative. Then, by the Campbell-Baker-Hausdorff formula, this gives

$$xy = \exp X \exp Y = \exp(X + Y) = \exp Y \exp X = yx. \qquad (4.19)$$

Therefore, W is commutative. But G is generated by W and so G is commutative.

Conversely, suppose that G is commutative. Then

$$\exp(tX)\exp(tY) = \exp\left(t\,(X + Y) + t^2\,[X, Y]/2 + O\left(t^3\right)\right) \qquad (4.20)$$

and

$$\exp(tY)\exp(tX) = \exp\left(t\,(Y + X) + t^2\,[Y, X]/2 + O\left(t^3\right)\right). \qquad (4.21)$$

Equating coefficients of t^2 gives

$$[X, Y] = [Y, X]. \qquad (4.22)$$

So, by the skew symmetry of $[\,,\,]$ we have $[X, Y] = 0$ and $\mathcal{G}$ is commutative.

We now study the situation when G is an abelian Lie group.

THEOREM 4.10   *If G is abelian $\exp$ is a homomorphism.*

Proof   We need to show $\exp X \exp Y = \exp(X + Y)$. But, this is an immediate consequence of the Campbell-Baker-Hausdorff formula.

LEMMA 4.11   *When G is abelian the set $\ker(\exp)$ is a discrete subgroup of $\mathcal{G}$.*

Proof   This follows since exp is a local diffeomorphism and a homomorphism. We can now give a classification of abelian Lie groups.

Theorem 4.12   *The only connected abelian Lie groups are $T^n \times \mathbf{R}^m$ where T^n is the n-dimensional torus. In particular, if G is an abelian compact Lie group, then G is a torus.*

Proof   We first note that in the connected abelian case exp is onto. Then $G \cong \mathcal{G}/\ker \exp$. As a group $\mathcal{G} \cong \mathbf{R}^k$ and $\ker \exp = \mathbf{Z}X_1 \oplus \ldots \oplus \mathbf{Z}X_n$. Thus $\mathcal{G}/\ker \exp \cong T^n \times \mathbf{R}^{k-n}$ and the result follows with $m = k - n$.

Exercises - Chapter 4

4.1   Let $A = \begin{pmatrix} ix & 0 \\ 0 & -ix \end{pmatrix}$ and $B = \begin{pmatrix} 0 & z \\ -\bar{z} & 0 \end{pmatrix}$. Use the result of Exercise 3.6 to calculate Z_1, Z_2 and Z_3 where $\exp(tA)\exp(tB) = \exp\left(tZ_1 + t^2 Z_2 + t^3 Z_3 + 0\left(t^4\right)\right)$.

4.2   From Theorem 4.1, deduce that if X and Y commute, that is $[X, Y] = 0$, then $\exp(X + Y) = \exp X \exp Y$.

4.3   Let the matrix X have the block form $X = \begin{pmatrix} A & 0 \\ 0 & B \end{pmatrix}$. Show that $\exp X$ has the block form $\begin{pmatrix} \exp(A) & 0 \\ 0 & \exp(B) \end{pmatrix}$.

4.4   Let $A \in \mathcal{SO}(2l + 1)$, the Lie algebra. Show that $\det A = 0$. Hence, show $A = P \begin{pmatrix} B & 0 \\ 0 & 0 \end{pmatrix} P^{-1}$ where $B \in \mathcal{SO}(2l)$, the Lie algebra. Using Exercise 4.3, and the fact that $\exp : \mathcal{SO}(n) \to SO(n)$ is onto, show that if $X \in SO(2l + 1)$, the group, then X is conjugate to a matrix $\begin{pmatrix} Y & 0 \\ 0 & 1 \end{pmatrix}$ with $Y \in SO(2l)$.

4.5   Let $\exp t(X+Y) = \exp tX \exp tY \exp W(t)$, so $W(t) = \sum_{n=1}^{\infty} t^n W_n$. Find W_1, W_2 and W_3.

4.6   Use the Campbell-Baker-Hausdorff formula to show that in $SU(2)$: $\exp\begin{pmatrix} ix & z \\ -\bar{z} & -ix \end{pmatrix}$
$$= \lim_{n\to\infty} \left(\exp\begin{pmatrix} ix/n & 0 \\ 0 & -ix/n \end{pmatrix} \exp\begin{pmatrix} 0 & z/n \\ -\bar{z}/n & 0 \end{pmatrix}\right)^n.$$

Now use Exercise 3.6 to try and show this directly.

4.7   Let $T = \left\{ \begin{pmatrix} e^{i\theta_1} & & \\ & \ddots & \\ & & e^{i\theta_n} \end{pmatrix} : \sum \theta_j = 0 \text{ and off diagonal elements} \right.$ are zero $\left. \right\}$. Show T is a closed subgroup of $SU(n)$.

4.8   Continue Exercise 4.7 and show, using Theorem 4.12, that T is a torus.

4.9   Let T be a torus in G with Lie algebra $\mathcal{T}$. Use Exercise 3.10 to show that the image of $\mathcal{T}$ under $\exp : \mathcal{T} \to G$ is T.

4.10   If T is a torus and is a subgroup of G show for any $g \in G$ that gTg^{-1} is also a torus subgroup of G.

Chapter 5

The Adjoint Representation

In this chapter, we discuss the adjoint representation. This represents a change of direction from the previous chapters. Eventually, we shall apply the work of previous chapters to the present material. Then, we shall obtain important results concerning the structure of the Lie group, its Lie algebra and its representations. We start by defining the adjoint representation, which is how the group acts on its Lie algebra.

Let $x \in G$. Then the map $I_x : G \to G$ is given by $I_x(g) = xgx^{-1}$. Clearly $I_x I_y = I_{xy}$.

DEFINITION 5.1 The adjoint representation of G is $Ad : G \to Aut\, \mathcal{G}$ defined by $Adx = (dI_x)_1$.

DEFINITION 5.2 The adjoint representation of $\mathcal{G}$ is $ad : \mathcal{G} \to End\, \mathcal{G}$ defined by $adX(Y) = [X, Y]$.

THEOREM 5.3 *These adjoint representations satisfy* $d(Ad)_1 = ad$.

Proof If f is any homomorphism $f : G \to H$ then

$$f \circ \exp_G = \exp_H \circ df_1, \tag{5.1}$$

where $\exp_G$ (respectively $\exp_H$) is the exponential on G (respectively H). Thus, if we take $I_{\exp(tX)}$ for f , which implies $G = H$, we get

$$I_{\exp(tX)} \circ \exp = \exp \circ Ad(\exp(tX)). \tag{5.2}$$

Applying this to the point tY in $\mathcal{G}$ gives

$$\exp(Ad(\exp(tX))(tY)) = \exp(tX)\exp(tY)\exp(-tX)$$
$$= \exp\left((tY) + t^2[X, Y] + O\left(t^3\right)\right). \tag{5.3}$$

33

Equation (5.3) follows from the Campbell-Baker-Hausdorff formula. For t small we get

$$Ad(\exp(tX))(Y) = Y + t\,[X,Y] + O\left(t^2\right). \tag{5.4}$$

Now we apply (5.1) again, but this time use the map $Ad : G \to Aut\,\mathcal{G}$ in place of f:

$$\exp_{Aut}\left(d(Ad)_1 tX\right)(Y) = Y + t\,[X,Y] + O\left(t^2\right). \tag{5.5}$$

Thus, the map $f : t \to Y + t\,[X,Y] + O\left(t^2\right)$ has the form

$$f(t) = \exp_{Aut}\left(tX_0\right)(Y), \tag{5.6}$$

where $X_0 = d(Ad)_1 X$. To finish the proof we carry out the following calculation:

$$
\begin{aligned}
X_0(Y) &= \frac{d}{dt}\left(\exp tX_0\right)|_{t=0} Y \\
&= \frac{d}{dt}\left(\exp tX_0\right) Y|_{t=0} \\
&= \frac{df}{dt}|_{t=0} \\
&= [X,Y].
\end{aligned} \tag{5.7}
$$

That is

$$d(Ad)_1(X)Y = [X,Y], \tag{5.8}$$

which completes the proof of Theorem 5.3.

Having defined the adjoint representation, we now proceed to use it. The first use is to define the Killing form on $\mathcal{G}$, and hence under certain conditions a Riemannian metric on G.

DEFINITION 5.4   The bilinear form $B(X,Y) = tr(adX\,adY)$ is called the Killing form on $\mathcal{G}$.

DEFINITION 5.5   If $<,>$ is a bilinear form on $\mathcal{G}$ then $<,>$ is invariant if it is invariant under Ad, that is, $< X,Y >=< Ad(g)(X), Ad(g)(Y) >$.

THEOREM 5.6   *The Killing form is invariant.*

Proof   Let $\sigma : \mathcal{G} \rightarrow \mathcal{G}$ be a Lie algebra automorphism. Then $ad(\sigma X) = \sigma \cdot adX \cdot \sigma^{-1}$. Now, we can calculate $B(Ad(g)X, Ad(g)Y)$ to prove the theorem:

$$
\begin{aligned}
B(Ad(g)X, Ad(g)Y) &= tr(ad(Ad(g)X)ad(Ad(g)Y)) \\
&= tr\left(Ad(g)adX\,adY(Ad(g))^{-1}\right) \\
&= tr(adX\,adY) \\
&= B(X,Y).
\end{aligned} \tag{5.9}
$$

Clearly, given the bilinear form on $\mathcal{G}$ we get a bilinear metric on G by left translation. This is, of course, left invariant. If the original bilinear form is invariant, then the resulting bilinear metric is also right invariant. This bilinear metric is a pseudo Riemannian metric, if the original bilinear form is non-degenerate. We get a Riemannian metric if the bilinear form is positive definite.

Definition 5.7   The group G is semi-simple, if B is non-degenerate.

The centre of $G, Z(G)$, is the set of all elements of G which commute with all other elements of G. That is $Z(G) = \{x \in G : xg = gx$ for all $g \in G\}$. Clearly $Z(G)$ is a closed subgroup of G and so, is a Lie subgroup. If $\mathcal{Z}(\mathcal{G})$ is the Lie algebra of $Z(G)$ then $adX = 0$ for $X \in \mathcal{Z}(g)$. Thus $B \mid \mathcal{Z}(\mathcal{G})$ is zero. Hence, if B is non-degenerate $\mathcal{Z}(\mathcal{G}) = 0$ and $Z(G)$ is discrete. The next theorem gives all the groups G for which B is definite.

Theorem 5.8   *A connected semi-simple group G is compact, if and only if B is negative definite.*

Proof   First, we prove that G compact implies that B is negative definite. Let $< , >$ be any inner product on $\mathcal{G}$. Then, we obtain an invariant inner product by integration:

$$
< X, Y >_I = \int < Ad(g)X, Ad(g)Y > dg. \tag{5.10}
$$

Here, we have used $< , >_I$ for the invariant inner product, the subscript I will now be suppressed. Since $< , >$ is invariant Adg is an orthogonal transformation of $\mathcal{G}$. Thus adX is skew symmetric, and so, if $adX = (A_{ij})$ relative to an orthonormal basis:

$$
\begin{aligned}
tr(adX\,adX) &= \sum_i \sum_k A_{ik}A_{ki} \\
&= -\sum_i \sum_k A_{ik}^2 \leq 0.
\end{aligned} \tag{5.11}
$$

Since G is semi-simple B is non-degenerate, and so, B is negative definite.

The converse result, that B negative definite implies G compact, requires a theorem due to Weyl. We quote this.

THEOREM   *(Weyl). If K is a compact connected group with a discrete centre, then the universal covering group, $\tilde{K}$, is also compact.*

The proof of this requires calculating the Ricci curvature for $\tilde{K}$ and will not be given here.

We now return to the proof of the theorem. Since B is negative definite Adg is orthogonal. Thus AdG is a subgroup of $O(n)$ and so, is compact. In the theorem of Weyl AdG plays the role of K. The universal covering group of AdG is also the universal covering group of G so, once we show that $Ker\,Ad$ is discrete, the proof is finished. Let $\mathcal{H}$ be the Lie algebra of $H = Ker\,Ad$. Then $ad\,|\,\mathcal{H} = 0$ and since B is negative definite $\mathcal{H} = 0$. That is, H is discrete.

It is a little unfortunate that the Killing form is negative definite, rather than positive definite. This is easily remedied. Two common conventions are: to work with the form $-B(X,Y)$ or to regard $\mathcal{G}$ as pure imaginary, that is, to work with $B(iX, iY)$. Each of these has its uses.

It is convenient to include here the definition of a nilpotent Lie algebra and group.

DEFINITION 5.9   A Lie algebra $\mathcal{G}$ is nilpotent if there is an integer k such that for all $X \in \mathcal{G}$ the endomorphism $(adX)^k = 0$. A group is call nilpotent if its Lie algebra is nilpotent.

Clearly, any abelian group is nilpotent. The other elementary example of a nilpotent group is the set of upper triangular matrices with diagonal entries all equal to 1. This group has for its Lie algebra the set of strictly upper triangular matrices.

Example 5.10   Let $G = SU(2)$. Then, since this is a matrix group, if $B \in \mathcal{SU}(2)$, the Lie algebra, then

$$\exp tB = I + tB + \dots. \tag{5.12}$$

Let $A \in SU(2)$ so $I_A \exp B = A \exp B A^{-1}$. Now since

$$dI_A B = \lim_{t \to 0} \frac{I_A \exp(tB) - I_A \exp(0)}{t}$$
$$= I_A B, \tag{5.13}$$

we see that $Ad : G \to Aut \, \mathcal{G}$ is given by $AdA = I_A$. Let $A = \begin{pmatrix} x + iy & u + iv \\ -u + iv & x - iy \end{pmatrix}$ and as a basis for the Lie algebra $SU(2)$ take

$$\begin{pmatrix} i & 0 \\ 0 & -i \end{pmatrix}, \begin{pmatrix} 0 & 1 \\ -1 & 0 \end{pmatrix} \text{ and } \begin{pmatrix} 0 & i \\ i & 0 \end{pmatrix}. \tag{5.14}$$

Then, a calculation gives

$$Ad \begin{pmatrix} x + iy & u + iv \\ -u + iv & x - iy \end{pmatrix} \tag{5.15}$$
$$= \begin{pmatrix} x^2 + y^2 - u^2 - v^2 & -2xv + 2uy & 2xu + 2yv \\ 2uy + 2xv & x^2 - y^2 + u^2 - v^2 & -2xy + 2uv \\ -2xu + 2yv & 2xy + 2uv & x^2 - y^2 - u^2 + v^2 \end{pmatrix}.$$

This takes a particularly simple and useful form on the diagonal elements of $SU(2)$:

$$Ad \begin{pmatrix} e^{i\theta} & 0 \\ 0 & e^{-i\theta} \end{pmatrix} = \begin{pmatrix} 1 & 0 & 0 \\ 0 & \cos 2\theta & -\sin 2\theta \\ 0 & \sin 2\theta & \cos 2\theta \end{pmatrix}. \tag{5.16}$$

Similarly, using the same basis for the Lie algebra, the adjoint representation of diagonal elements of the Lie algebra is given by

$$ad \begin{pmatrix} i\theta & 0 \\ 0 & -i\theta \end{pmatrix} = \begin{pmatrix} 0 & 0 & 0 \\ 0 & 0 & -2\theta \\ 0 & 2\theta & 0 \end{pmatrix}. \tag{5.17}$$

A simple calculation now gives

$$tr \, ad \begin{pmatrix} i\theta & 0 \\ 0 & -i\theta \end{pmatrix} ad \begin{pmatrix} i\varphi & 0 \\ 0 & -i\varphi \end{pmatrix} = -8\,\theta\,\varphi. \tag{5.18}$$

Example 5.11   Using $SU(2)$ as a guide, we see that in general the expression for the adjoint representation of the groups $U(n), SU(n)$ and $Sp(n)$ are too complicated to be enlightening. However, if we restrict our attention to the diagonal elements of these groups, we can obtain the expressions for the Killing forms. As we shall see in the next chapter, it is, in fact, these expressions which are the most useful ones. In all three cases, we work with the Lie algebra rather than with the group.

a)   The Unitary groups $U(n)$.

We use the following basis for the Lie algebra $\mathcal{U}(n)$:

D_j =   diagonal matrix with i in the jth entry,
E_{ij} =   matrix with 1 in the (i,j)th entry and -1 in the (j,i)th entry,
F_{ij} =   matrix with $\mathbf{i}$ in both the (i,j)th and (j,i)th entries.   (5.19)

In all cases, the matrices D_j, E_{ij}, F_{ij} have zeros in the other entries and E_{ij} and F_{ij} are only considered for $i < j$. To avoid confusion and distinguish between the index i and $\sqrt{-1}$ we use boldface $\mathbf{i} = \sqrt{-1}$. Let

$$A = \begin{pmatrix} \mathbf{i}\,\theta_1 & & \\ & \ddots & \\ & & \mathbf{i}\,\theta_n \end{pmatrix}.$$ (5.20)

Then a simple calculation gives

$$\begin{aligned} AD_j - D_jA &= 0 \\ AE_{ij} - E_{ij}A &= (\theta_i - \theta_j)\,F_{ij} \\ AF_{ij} - F_{ij}A &= (\theta_j - \theta_i)\,E_{ij}. \end{aligned}$$ (5.21)

Now if

$$B = \begin{pmatrix} \mathbf{i}\,\varphi_1 & & \\ & \ddots & \\ & & \mathbf{i}\,\varphi_n \end{pmatrix},$$ (5.22)

then a simple calculation gives

$$tr\ ad\ A\ ad\ B = -2n \sum_{i=1}^{n} \theta_i\,\varphi_i + 2 \left(\sum_{i=1}^{n} \theta_i \right) \left(\sum_{i=1}^{n} \varphi_i \right).$$ (5.23)

Notice that when all $\theta_i = 1$ and all $\varphi_i = 1$ this is zero; that is

$$tr\ ad\ (\mathbf{i}\,I)\ ad(\mathbf{i}\,I) = 0.$$ (5.24)

Thus, we see that the Killing form on $U(n)$ is not negative definite, which is correct since $U(n)$ is not semi- simple.

b)   The special Unitary group $SU(n)$.

This is very similar to the case of the Unitary group $U(n)$. We make the following changes:

$D_j =$ diagonal matrix with $\mathbf{i}$ in the jth place and $-\mathbf{i}$ in the nth place, (5.25)

and in equations (5.20) and (5.22), we impose the conditions

$$\sum_{i=1}^{n} \theta_i = 0 \text{ and } \sum_{i=1}^{n} \varphi_i = 0. \tag{5.26}$$

Again, we use boldface to distinguish between $\mathbf{i} = \sqrt{-1}$ and the index i. The calculations now proceed as before yielding the result

$$tr\, ad\, A\, ad\, B = -2n \sum_{i=1}^{n} \theta_i\, \varphi_i. \tag{5.27}$$

This, together with the results of the next chapter, will give that the Killing form is negative definite on $SU(n)$. Again, this is correct since $SU(n)$ is semi-simple.

c)   The symplectic group $Sp(n)$.

In this case, we shall only be interested in the diagonal matrices with complex entries, rather than quaternionic entries. The reason for this will become clear in the next chapter. This case is sufficiently similar to that of $U(n)$ that we use A, B, D_j, E_{ij} and F_{ij} as in part a). In addition, we define

$$G_{ij} = \text{matrix with } \mathbf{j} \text{ in the } (i,j)\text{th and } -\mathbf{j} \text{ in the } (j,i)\text{th entries,}$$
$$H_{ij} = \text{matrix with } \mathbf{k} \text{ in the } (i,j)\text{th and } -\mathbf{k} \text{ in the } (j,i)\text{th entries,}$$
$$K_i = \text{diagonal matrix with } \mathbf{j} \text{ in the } i\text{th place,}$$
$$\text{and } L_i = \text{diagonal matrix with } \mathbf{k} \text{ in the } i\text{th place.} \tag{5.28}$$

Where we have used boldface $\mathbf{i}, \mathbf{j}, \mathbf{k}$ to denote the unit quaternions and distinguish these from the same letter used as an index. As for the unitary group, we now calculate:

$$AD_j - D_j A = 0,$$
$$AE_{ij} - E_{ij}A = (\theta_i - \theta_j)\, F_{ij},$$
$$AF_{ij} - F_{ij}A = (\theta_j - \theta_i)\, E_{ij},$$

$$
\begin{aligned}
AG_{ij} - G_{ij}A &= (\theta_i + \theta_j)\, H_{ij}, \\
AH_{ij} - H_{ij}A &= -(\theta_i + \theta_j)\, G_{ij}, \\
AK_i - K_i A &= 2\theta_i L_i, \\
AL_i - L_i A &= -2\theta_i K_i.
\end{aligned} \tag{5.29}
$$

The calculation of the Killing form now proceeds in a familiar fashion to give

$$
tr\ ad\ A\ ad\ B = -4(n+1) \sum_{i=1}^{n} \theta_i\, \varphi_i. \tag{5.30}
$$

Example 5.12   The special orthogonal groups $SO(n)$ do not have any interesting diagonal elements – just the matrices with ± 1 on the diagonal with an even number of -1. As we shall see in the next chapter, the diagonal matrices need to be replaced by ones with 2×2 rotation matrices as blocks on the diagonal. These blocks are

$$
\begin{pmatrix} \cos\theta & \sin\theta \\ -\sin\theta & \cos\theta \end{pmatrix}. \tag{5.31}
$$

We use the notation

$$
\Theta_i = \begin{pmatrix} 0 & \theta_i \\ -\theta_i & 0 \end{pmatrix} \tag{5.32}
$$

for an element of the Lie algebra of $SO(2)$ which, if $\theta = \theta_i$, maps to the matrix (5.31) under the exponential map. There are two cases: when n is odd and n is even.

   a)   The groups $SO(2l + 1)$.

As a basis for the Lie algebra $SO(2l + 1)$ we take the matrices

$E_{i,j}$ = matrix with   1   in the (i,j)th entry, and -1 in the (j,i)th entry. $\tag{5.33}$

this is for $i < j$ and there are zeros in all other entries. Let

$$
A = \begin{pmatrix} \Theta_1 & & \\ & \ddots & \\ & & \Theta_l \\ & & & 0 \end{pmatrix}. \tag{5.34}
$$

Then a calculation gives the following cases

$$
AE_{ij} - E_{ij}A = \theta_{\frac{i}{2}} E_{i-1,j} + \theta_{\frac{j}{2}} E_{i,j-1} \quad i,j \text{ both even} \tag{5.35}
$$

$$= \theta_{\frac{i}{2}} E_{i-1,j} - \theta_{\frac{i+1}{2}} E_{i,j+1} \quad i \text{ even}, j \text{ odd} \quad (5.36)$$

$$= -\theta_{\frac{i+1}{2}} E_{i+1,j} + \theta_{\frac{i}{2}} E_{i,j-1} \quad i \text{ odd}, j \text{ even} \quad (5.37)$$

$$= -\theta_{\frac{i+1}{2}} E_{i+1,j} - \theta_{\frac{i+1}{2}} E_{i,j+1} \quad i, j \text{ both odd}, (5.38)$$

where $\theta_k = 0$ if $k > l$ and $E_{ij} = 0$ unless $i < j \le 2l + 1$. Notice that a comma has been inserted for clarity. That is $E_{i,j} = E_{ij}$ with the comma indicating the point of separation between the subscripts when either of them is given by a lengthy expression. If

$$B = \begin{pmatrix} \Phi_1 & & \\ & \ddots & \\ & & \Phi_l \ 0 \end{pmatrix}, \qquad (5.39)$$

then using (5.35) we find

$$tr \, ad \, A \, ad \, B = -(4 \, l - 2) \sum_{k=1}^{l} \theta_k \varphi_k. \qquad (5.40)$$

b)   The groups $SO(2 \, l)$.

Since these are similar to the groups $SO(2 \, l + 1)$, we use the same notation for the basis E_{ij} of the Lie algebra. Notice this time that

$$A = \begin{pmatrix} \Theta_1 & & \\ & \ddots & \\ & & \Theta_l \end{pmatrix}, \qquad (5.41)$$

with the zero in the last entry omitted. Again, we find

$$AE_{ij} - E_{ij}A = \theta_{\frac{i}{2}} E_{i-1,j} + \theta_{\frac{j}{2}} E_{i,j-1} \quad i, j \text{ both even} \quad (5.42)$$

$$= \theta_{\frac{i}{2}} E_{i-1,j} - \theta_{\frac{i+1}{2}} E_{i,j+1} \quad i \text{ even}, j \text{ odd} \quad (5.43)$$

$$= -\theta_{\frac{i+1}{2}} E_{i+1,j} + \theta_{\frac{j}{2}} E_{i,j-1} \quad i \text{ odd}, j \text{ even} \quad (5.44)$$

$$= -\theta_{\frac{i+1}{2}} E_{i+1,j} - \theta_{\frac{i+1}{2}} E_{i,j+1} \quad i, j \text{ both odd}, (5.45)$$

where $\theta_k = 0$ if $k > l$ and $E_{ij} = 0$ unless $i < j \le 2l$. Then, we find

$$tr \, ad \, A \, ad \, B = -(4 \, l - 4) \sum_{k=1}^{l} \theta_k \varphi_k. \qquad (5.46)$$

Exercises - Chapter 5

5.1   In $SO(3)$ let $A = \begin{pmatrix} 0 & \theta & 0 \\ -\theta & 0 & 0 \\ 0 & 0 & 0 \end{pmatrix}$, $E_{12} = \begin{pmatrix} 0 & 1 & 0 \\ -1 & 0 & 0 \\ 0 & 0 & 0 \end{pmatrix}$, $E_{13} = \begin{pmatrix} 0 & 0 & 1 \\ 0 & 0 & 0 \\ -1 & 0 & 0 \end{pmatrix}$ and $E_{23} = \begin{pmatrix} 0 & 0 & 0 \\ 0 & 0 & 1 \\ 0 & -1 & 0 \end{pmatrix}$.   Show

  a) $ad(A)E_{12} = 0$;
  b) $ad(A)E_{13} = -\theta E_{23}$; and
  c) $ad(A) E_{23} = \theta E_{13}$.

5.2   Show that the result of Exercise 5.1 agrees with equation (5.35) in the case of $SO(3)$.

5.3   Let G be an abelian group. Show the Killing form is zero: $B(X, Y) = 0$ for all X and Y.

5.4   Let Z be the center of G and suppose that Z is not discrete. If $\mathcal{Z}$ is the Lie algebra of Z show that for $X \in \mathcal{Z}$ and $Y \in \mathcal{G}$ that $B(X, Y) = 0$.

5.5   For the group $U(n)$ let X be a diagonal matrix in the kernel of the Killing form, that is, $tr\, ad\, X\, ad\, Y = 0$ for all Y. Use equation (5.23) to show that $X = i\theta I$ and so $ad\, X = 0$.

5.6   Use the results of Exercises 5.4 and 5.5 to show that the only diagonal matrices in the center of $U(n)$ are $e^{i\theta} I$.

5.7   Let G be a compact semi-simple Lie group with center Z. Show that Z must be finite.

5.8   By a direct calculation, show that the center of $SU(2)$ is $\{I, -I\}$ where $I = \begin{pmatrix} 1 & 0 \\ 0 & 1 \end{pmatrix}$.

5.9   Show that the center, Z, of a connected compact Lie group is $Z = \ker Ad$.

5.10   If G is a matrix group, notice that $A \in \mathcal{G}$ is also given by a matrix. Show that $(d\, I_g)_1 A = g A g^{-1}$ and so, notice that for a matrix group $(Ad\, g)A = g A g^{-1}$.

Chapter 6

Maximal Tori

In this chapter, we study maximal tori in a Lie group. The basic result is that any two maximal tori are conjugates. It is also true that a connected Lie group is the union of its maximal tori. While this fact is important, it is a trivial consequence of our definitions. We start with two definitions.

DEFINITION 6.1   Let G be a Lie group and $g \epsilon G$. If H is the subgroup generated by g then g is called a generator of G if $\overline{H} = G$.

DEFINITION 6.2   The group G is called monogenic, if it has a generator.

PROPOSITION 6.3   *The torus T is monogenic.*

Proof   Let $T = \mathbf{R}^k / \mathbf{Z}^k$ have coordinates $x_1, \ldots, x_k$ and let $U_1, U_2, \ldots,$ be a basis for open sets. A cube in T is a set

$$C = \{x \epsilon T : | \; x_i - \xi_i \, | \le \epsilon \} \tag{6.1}$$

for some fixed $\xi \in T$ and real $\epsilon > 0$. The number 2ϵ is called the side of C. We shall define a descending sequence of cubes:

$$C_0 \supset C_1 \supset \ldots C_{m-1} \supset \ldots . \tag{6.2}$$

Suppose that cube C_{m-1} has side $2 \epsilon_{m-1}$. Then, there is a positive integer n_{m-1} such that

$$n_{m-1} 2 \epsilon_{m-1} > 1.$$

Thus, the image of C_{m-1} under multiplication by n_{m-1} is T. Let C_m be a cube in C_{m-1} such that $n_{m-1} C_m \subset U_m$. Since C_m is closed, the intersection, $\cap C_m$, is non-empty. Let $g \in \cap C_m$. Then $g^{n_{m-1}} \in U_m$ for all m. That is: g is a generator of T.

We now define a maximal torus .

DEFINITION 6.4   A torus T is maximal if $T \subset U \subset G$ and U is a torus then $T = U$.

THEOREM 6.5   *Any torus is contained in a maximal torus.*

Proof   Consider an increasing sequence of tori:

$$T \subset T_1 \subset T_2 \subset \ldots G. \tag{6.3}$$

Then, we must show that this is finite. Take Lie algebras of this chain to get

$$T(T) \subset T(T_1) \subset \ldots \subset T(G). \tag{6.4}$$

Since this is an increasing sequence of finite dimensional vector spaces, it must be a finite sequence.

It turns out that maximal tori in a connected group are also maximal abelian subgroups. At the moment, we cannot prove this. However, we can prove the following weaker result.

THEOREM 6.6   *If T is a maximal torus, then T is a maximal connected abelian subgroup of G.*

Proof   If $T \subset A$ and A is connected an abelian, then $T \subset \bar{A}$. Now $\bar{A}$ is closed and G is compact, so $\bar{A}$ is compact. However, a compact, connected abelian group is a torus. Since T is a maximal torus, we have $T = \bar{A}$, which completes the proof.

Notice that Theorem 6.6 holds even when G is not connected. However, when G is connected, any maximal abelian subgroup of G, which contains a maximal torus, is also connected. When G is not connected, then a maximal abelian subgroup need not be a torus. The group $O(2n+1)$ illustrates this since $-I$ is in the center of $O(2n+1)$, and hence, in any maximal abelian subgroup. On the other hand $-I$ is not in the same connected component as the identity, and therefore, $-I$ is not in any maximal torus of $O(2n+1)$.

The two preceding results indicate a useful line to follow. In Theorem 6.5 we were able to obtain information about the group from its Lie algebra. Theorem 6.6 shows that the notion of a maximal connected abelian subgroup can be used in place of that of a maximal torus. The following result lets us exploit these ideas.

THEOREM 6.7 *A connected immersed subgroup $H \subset G$ is a maximal torus of G if its Lie algebra $\mathcal{H}$ is a maximal commutative Lie subalgebra of $\mathcal{G}$.*

Proof We need only show that H is closed in G. Now $\bar{H}$ is a connected, abelian, Lie subgroup of G. Thus $T_1(\bar{H})$ is an abelian subalgebra of $\mathcal{G}$ with $\mathcal{H} \subset T_1(\bar{H})$. By the maximality of $\mathcal{H}$ we have $\mathcal{H} = T_1(\bar{H})$ so $H = \bar{H}$.

To proceed further, we introduce two subspaces of $\mathcal{G}$. These are $\mathcal{G}^X$ and $\mathcal{G}_X$.

DEFINITION 6.8 If $X \in \mathcal{G}$ then let

$$\mathcal{G}_X = \{Y \in \mathcal{G} : [X, Y] = 0\}$$

and

$$\mathcal{G}^X = \{Z \in \mathcal{G} : Z = ad(X)Y \quad \text{for some} \quad Y \in \mathcal{G}\}.$$

The second of these spaces is the image of $\mathcal{G}$ under $ad(X)$, that is $\mathcal{G}^X = ad(X)\mathcal{G}$. There is a subgroup G_X corresponding to $\mathcal{G}_X$. This correspondence is given by

$$\mathcal{G}_X = T_1(G_X) \tag{6.5}$$

and

$$G_X = \{g \in G : Ad(g)X = 0\}. \tag{6.6}$$

LEMMA 6.9 *With respect to a positive definite Ad-invariant inner-product, the spaces $\mathcal{G}_X$ and $\mathcal{G}^X$ are perpendicular.*

Proof Let $Y \in \mathcal{G}$. Then, since $<,>$ is Ad invariant, we have

$$< Ad(exp\,tZ)X, Ad(exp\,tZ)Y > \; = \; < X, Y >, \tag{6.7}$$

for any X and $Y \in \mathcal{G}$. Now, differentiate this with respect to t and set $t = 0$ to get

$$< [Z, X], Y > + < X, [Z, Y] > = 0. \tag{6.8}$$

Now, let $Y \in (\mathcal{G}^X)^\perp$ the space perpendicular to $\mathcal{G}^X$. Since $[Z, X] \in \mathcal{G}^X$, we have $< [Z, X], Y > = 0$, and hence, $< X, [Z, Y] > = 0$. Thus, using (6.8) with Y and Z interchanged gives

$$< [X, Y], Z > = 0. \tag{6.9}$$

Since this holds for all Z and the innerproduct is nondegenerate, we have $[X, Y] = 0$, and therefore, $Y \in \mathcal{G}_X$. Thus $(\mathcal{G}^X)^\perp \subset \mathcal{G}_X$.

Now, we count dimensions to show $(\mathcal{G}^X)^\perp = \mathcal{G}_X$. First, notice $\dim \mathcal{G}_X = \dim \ker ad\, X$ and $\dim \mathcal{G}^X = \dim Im\, ad\, X$. Thus, $\dim \mathcal{G}_X + \dim \mathcal{G}^X = \dim \mathcal{G}$. Now, $\dim \mathcal{G}^X + \dim(\mathcal{G}^X)^\perp = \dim \mathcal{G}$ since the innerproduct is nondegenerate, and therefore, $\dim \mathcal{G}_X = \dim(\mathcal{G}^X)^\perp$ which completes the proof.

It is now of interest to show when we have or do not have an *Ad*-invariant innerproduct.

If G is either semisimple or abelian, then $\mathcal{G}$ has a positive definite *Ad*-invariant innerproduct. In the case of a semisimple group, this is the Killing form. For an abelian group, notice that the connected component of the identity is R^n/L for some lattice L. Now, take the standard innerproduct on R^n.

On the other hand, a non-abelian nilpotent group does not have a positive definite innerproduct. Let $X, Y \in \mathcal{G}$ for such a group with

$$(ad\, X)^k Y = 0 \text{ and } (ad\, X)^{k-1} Y \neq 0 \tag{6.10}$$

for some $k \geq 2$. Let $Z = (ad\, X)^{k-2} Y$. Then, we see

$$[X, [X, Z]] = 0 \text{ and } [X, Z] \neq 0. \tag{6.11}$$

Now, by (6.8)

$$< [X, Z], [X, Z] > + < [X, [X, Z]], Z > = 0, \tag{6.12}$$

and therefore,

$$< [X, Z], [X, Z] > = 0, \tag{6.13}$$

and the innerproduct is not positive definite.

When G is compact, we can use integration to turn any innerproduct into an *Ad*-invariant one. In the more general setting of any representation, rather than just the adjoint representation, this is the content of Theorem 7.6. Notice, of course, that a non-abelian nilpotent group is not compact.

DEFINITION 6.10   An element $X \in \mathcal{G}$ is regular, if $\dim \mathcal{G}_X \leq \dim \mathcal{G}_Y$ for all $Y \in \mathcal{G}$.

LEMMA 6.11   *If X is regular, then $\mathcal{G}_X$ is abelian.*

Proof   We suppose that $\mathcal{G}_X$ is not abelian. Then, we can find $Y, Z \in \mathcal{G}_X$ such that $[Y, Z] \neq 0$. We shall consider $\mathcal{G}_{X+tY}$. Now, $[X+tY, X] = 0$ and so, $ad(X + tY)$ preserves the decomposition

$$\mathcal{G} = \mathcal{G}_X \oplus \mathcal{G}^X. \tag{6.14}$$

But, $ad\, X$ acts isomorphically on $\mathcal{G}^X$ and therefore, for t sufficiently small $ad(X + tY)$ also acts isomorphically. That is,

$$\mathcal{G}^X \subset \mathcal{G}^{X+tY}, \tag{6.15}$$

and hence,

$$\mathcal{G}_{X+tY} \subset \mathcal{G}_X. \tag{6.16}$$

However, $Z \notin \mathcal{G}_{X+tY}$ and therefore, $\dim \mathcal{G}_{X+tY} < \dim \mathcal{G}_X$ with a strict inequality. This last fact contradicts X being regular, and so, $\mathcal{G}_X$ is abelian.

DEFINITION 6.12   When X is regular $\mathcal{G}_X$ is called a Cartan subalgebra.

Our basic problem is to show that any two maximal tori are conjugate. We can now prove that any point in the Lie algebra is conjugate to a point in a given Cartan subalgebra. This is a major step towards our goal.

THEOREM 6.13   *Let H be a Cartan subalgebra and $Y \in \mathcal{G}$. Then, $(Ad\, G)Y \cap H$ is a finite, non-empty set of points.*

Proof   Let $H = \mathcal{G}_X$ and define a real valued function f on the orbit $(Ad\, G)Y$ by

$$f(Z) = <Z, X>. \tag{6.17}$$

Since $(Ad\, G)Y$ is compact f attains its minimum. We can suppose that this minimum is attained at Y. Now, let $Z \in \mathcal{G}$. Then,

$$\frac{d}{dt} f(Ad(\exp tZ)Y) \mid_{t=0} = 0 \tag{6.18}$$

since Y is a minimum of f. Thus,

$$<[Z, Y], X> = 0 \quad \text{for all} \quad Z \in \mathcal{G}. \tag{6.19}$$

Therefore, we have

$$< Y, [\mathcal{G}, X] > \; = 0. \tag{6.20}$$

Equation (6.20) says that $Y \in (\mathcal{G}^X)^{\perp}$, that is, $Y \in \mathcal{G}_X$. Thus, we have shown that $(Ad\, G)Y \cap H$ is non-empty.

Let U be a neighborhood of 1 in G such that $\exp : V \to U$ is a diffeomorphism onto U. Then, consider a curve $\gamma(t) = \exp tv$ in U, for some $v \in V$, and therefore,

$$Ad(\gamma(t))Y = Ad(\exp tv)Y. \tag{6.21}$$

Thus, the tangent vector to this curve is

$$[v, Y] \epsilon \, \mathcal{G}^Y, \tag{6.22}$$

which is perpendicular to $\mathcal{G}_Y$,and hence, to $\mathcal{G}_X$, since $\mathcal{G}_Y \supset \mathcal{G}_X$ because $Y \in \mathcal{G}_X$. Thus, there exists U , and open neighborhood of 1 in G, such that

$$Ad(U)Y \cap H = \{Y\}. \tag{6.23}$$

Now, the compactness of G implies $Ad(G)Y \cap H$ is finite.

Lemma 6.14   *If G is connected, then the exponential map is onto.*

Proof   Let $g \in G$. Since G is compact, we may join g to 1 by a geodesic. This is a consequence of the Hopf-Rinow theorem. Near 1, this geodesic has the form $\exp tX$. So, for some t, we have $\exp tX = g$, and so, $\exp$ is onto.

Definition 6.15   A connected subgroup H of G is a Cartan subgroup, if its Lie algebra is a Cartan subalgebra.

Theorem 6.16   *If T is a maximal torus and H is a Cartan subgroup, then there is $y \in G$ such that $y^{-1}Hy = T$.*

Proof   Let u be a generator for T and $Y \in \mathcal{G}$ such that $\exp Y = u$. Then, there is $y \in G$ such that

$$Ad(y)u \, \epsilon \, \mathcal{G}_X, \tag{6.24}$$

where $\mathcal{G}_X$ is the Lie algebra of $\tilde{H}$. Thus, we have

$$y\,uy^{-1} \in H, \tag{6.25}$$

therefore,

$$y\, Ty^{-1} \subset H. \tag{6.26}$$

Now H is a connected abelian subgroup of G so, by the maximality of T we have $y\, Ty^{-1} = H$ or what is equivalent to $T = y^{-1}Hy$.

COROLLARY 6.17 *If G is connected, then any element of G is conjugate to an element in T.*

COROLLARY 6.18 *In a connected group, any two maximal tori are conjugate.*

We now give the maximal tori for the classical groups. In the previous chapter, we were able to calculate the adjoint representation and the Killing form, not on the whole group and Lie algebra, but, only on a subgroup. It turns out that the subgroups which worked are the maximal tori of the groups concerned. From Corollary 6.11, we know that any element is conjugate to one in the maximal torus. Hence, in principle, the calculations at the end of the previous chapter are sufficient to deduce the adjoint representation and Killing form on the whole Lie algebra.

Example 6.19 The maximal torus of $SU(2)$. Set

$$T = \left\{ \begin{pmatrix} e^{i\theta} & 0 \\ 0 & e^{-i\theta} \end{pmatrix} : \theta \in \mathbf{R} \right\}. \tag{6.27}$$

Then, clearly, T is a torus inside $SU(2)$, in fact, T is a circle which is a one-dimensional torus. To show that T is maximal, let

$$A = \begin{pmatrix} z & w \\ -\bar{w} & \bar{z} \end{pmatrix} \tag{6.28}$$

be an element of $SU(2)$ which commutes with every element of T. Then

$$\begin{pmatrix} e^{i\theta} & 0 \\ 0 & e^{-i\theta} \end{pmatrix} \begin{pmatrix} z & w \\ -\bar{w} & \bar{z} \end{pmatrix} = \begin{pmatrix} e^{i\theta}z & e^{i\theta}w \\ e^{-i\theta}\bar{w} & e^{-i\theta}\bar{z} \end{pmatrix} \tag{6.29}$$

and

$$\begin{pmatrix} z & w \\ -\bar{w} & \bar{z} \end{pmatrix} \begin{pmatrix} e^{i\theta} & 0 \\ 0 & e^{-i\theta} \end{pmatrix} = \begin{pmatrix} e^{i\theta}z & e^{-i\theta}w \\ -e^{i\theta}\bar{w} & e^{-i\theta}\bar{z} \end{pmatrix}. \tag{6.30}$$

Thus, we have $e^{i\theta}w = e^{-i\theta}w$ for all values of θ. This is only possible if $w = 0$. Then, since $z\bar{z} + w\bar{w} = 1$, we have $z\bar{z} = 1$, i.e., $z = e^{i\phi}$ for some ϕ. This means that

$$A = \begin{pmatrix} e^{i\phi} & 0 \\ 0 & e^{-i\phi} \end{pmatrix} \qquad (6.31)$$

or that $A \in T$. We conclude that T is a maximal abelian subgroup of $SU(2)$.

Example 6.20 The maximal torus of $SO(3)$. This time set

$$T = \left\{ \begin{pmatrix} \cos\theta & \sin\theta & 0 \\ -\sin\theta & \cos\theta & 0 \\ 0 & 0 & 1 \end{pmatrix} \right\}, \qquad (6.32)$$

which is clearly a circle group, and so, a torus. Let A be an element of $SO(3)$, which commutes with T. Then, if $e_3 = \begin{pmatrix} 0 \\ 0 \\ 1 \end{pmatrix}$ and $t \in T$, we have

$$tAe_3 = Ate_3 = Ae_3. \qquad (6.33)$$

Thus, Ae_3 is an eigenvector for every element of T, that is, $Ae_3 = \lambda e_3$. Since A is in $SO(3)$, $\lambda = \pm 1$, and so, A has the form

$$A = \begin{pmatrix} B & 0 \\ & & 0 \\ 0 & 0 & 1 \end{pmatrix}, \qquad (6.34)$$

where B is a 2×2 block matrix. Since $A \in SO(3)$, we have for $\lambda = 1$

$$B = \begin{pmatrix} \cos\phi & \sin\phi \\ -\sin\phi & \cos\phi \end{pmatrix}. \qquad (6.35)$$

Thus, $A \in T$. On the other hand if $\lambda = -1$ the eigenvalues of A are $-1, -1$ and 1. Thus

$$A = \begin{pmatrix} -1 & & \\ & 1 & \\ & & -1 \end{pmatrix} \quad \text{or} \quad A = \begin{pmatrix} 1 & & \\ & -1 & \\ & & -1 \end{pmatrix},$$

neither of which commute with T.

Example 6.21 The groups $U(n)$, $SU(n)$, $SO(n)$ and $Sp(n)$.

a) The maximal torus of $U(n)$ is

$$T = \left\{ \begin{pmatrix} e^{i\theta_1} & & \\ & \ddots & \\ & & e^{i\theta_n} \end{pmatrix} \right\}. \tag{6.36}$$

First, we observe that T is a torus. Then, we set

$T_j =$ subgroup of T consisting of matrices with 1 in jth diagonal entry. $\tag{6.37}$

Let A be an element of $SU(n)$ which commutes with T. Then, if $t_j \in T_j$ we see

$$t_j A e_j = A t_j e_j = A e_j. \tag{6.38}$$

Thus, $A e_j$ is left fixed by T_j and so $A e_j = \lambda_j e_j$. Since $A \in U(n)$ all its eigenvalues are complex numbers of absolute value 1 so $\lambda_j = e^{i\phi_j}$ for some ϕ_j. Thus,

$$A = \begin{pmatrix} e^{i\phi_1} & & \\ & \ddots & \\ & & e^{i\phi_n} \end{pmatrix}, \tag{6.39}$$

and therefore, $A \in T$, that is, T is a maximal abelian subgroup of $U(n)$.

b) The maximal torus of $SU(n)$ is

$$T = \left\{ \begin{pmatrix} e^{i\theta_1} & & \\ & \ddots & \\ & & e^{i\theta_n} \end{pmatrix} : \theta_1 + \ldots + \theta_n = 0 \right\}. \tag{6.40}$$

To see that this is a torus, we use the isomorphism given by

$$\begin{pmatrix} e^{i\theta_1} & & \\ & \ddots & \\ & & e^{i\theta_n} \end{pmatrix} \rightarrow \begin{pmatrix} e^{i(\theta_1 - \theta_n)} & & \\ & \ddots & \\ & & e^{i(\theta_{n-1} - \theta_n)} \end{pmatrix}, \tag{6.41}$$

which maps T onto the maximal torus in $U(n-1)$. The argument given for $U(n)$ now shows that T is a maximal abelian subgroup in the case $n \geq 3$. The group $SU(2)$ has already been treated as a special case.

c)   That maximal torus of $SO(2n+1)$ is

$$
T = \left\{ \begin{pmatrix}
\begin{matrix} \cos\theta_1 & \sin\theta_1 \\ -\sin\theta_1 & \cos\theta_1 \end{matrix} & & & \\
& \ddots & & \\
& & \begin{matrix} \cos\theta_n & \sin\theta_n \\ -\sin\theta_n & \cos\theta_n \end{matrix} & \\
& & & 1
\end{pmatrix} \right\}. \tag{6.42}
$$

As before this is clearly a torus and so it only remains to see that it is a maximal. Let $A \in SO(2n+1)$ which commutes with T and let V_i be the subspace of $\mathbf{R}^{2n+1}$ spanned by e_{2i-1} and e_{2i}. Then T leaves each V_i invariant and there are subgroups T_i of T which leave V_i fixed. Notice that the whole of T leaves e_{2n+1} fixed. Now, we calculate with $t \in T$:

$$
tAe_{2n+1} = Ate_{2n+1} = Ae_{2n+1}. \tag{6.43}
$$

So, since λe_{2n+1} is the only vector fixed by the whole of T we have $Ae_{2n+1} = \lambda e_{2n+1}$. Since $A \in SO(2n+1)$ its only real eigenvalues are ± 1 and so $Ae_{2n+1} = \pm e_{2n+1}$. Now, let $t_i \in T_i$ and $v_i \in V_i$. Then,

$$
t_i Av_i = At_i v_i = Av_i. \tag{6.44}
$$

Thus Av_i is fixed by t_i and $Av_i \neq \lambda e_{2n+1}$. Hence $Av_i \in V_i$ that is, each space V_i is invariant under A. Since $A \in SO(2n+1)$ this means that $A \in T$.

d)   The maximal torus of $SO(2n)$ is

$$
T = \left\{ \begin{pmatrix}
\begin{matrix} \cos\theta_1 & \sin\theta_1 \\ -\sin\theta_1 & \cos\theta_1 \end{matrix} & & \\
& \ddots & \\
& & \begin{matrix} \cos\theta_n & \sin\theta_n \\ -\sin\theta_n & \cos\theta_n \end{matrix}
\end{pmatrix} \right\}. \tag{6.45}
$$

Consider the injection $SO(2n) \to SO(2n+1)$ given by

$$
A \to \begin{pmatrix}
A & \begin{matrix} 0 \\ \vdots \\ 0 \end{matrix} \\
0 \cdots 0 & 1
\end{pmatrix}. \tag{6.46}
$$

Then T is the inverse image of the maximal torus of $SO(2n+1)$, and so, it is a maximal torus.

e)   The maximal torus of $Sp(n)$ is

$$T = \left\{ \begin{pmatrix} e^{i\theta_1} & & \\ & \ddots & \\ & & e^{i\theta_n} \end{pmatrix} \right\}. \tag{6.47}$$

Notice that this is the same as the maximal torus of $U(n)$ which is a subgroup of $Sp(n)$. Let $A \in Sp(n)$, which commutes with T. Then, by the argument of part a) we have $Ae_j = \lambda_j e_j$. Now λ_j is a quaternion of unit length. Since A commutes with T we have $i\lambda_j = \lambda_j i$ so if $\lambda_j = a + bi + cj + dk$ we have

$$ia - b + ck - dj = ai - b - ck + dj \tag{6.48}$$

and so $c = d = 0$ and λ_j is complex. Thus T is maximal abelian in $Sp(n)$.

It is straightforward to show that $U(n)$ and $SU(n)$ are connected. See Exercises 6.6 and 6.7.

Example 6.22 $SO(n)$ is connected and $O(n)$ has two connected components.

To prove that $SO(n)$ is connected, we start with a result on real symmetric matrices.

LEMMA 6.23   *If S is a real symmetric matrix, then S has a real eigenvalue.*

Proof   By the fundamental theorem of algebra, S has at least one possibly complex eigenvalue λ :

$$Sv = \lambda v, \ \lambda \in \mathbf{c}, \ v \in \mathbf{c}^n. \tag{6.49}$$

Now, the standard Hermitian innerproduct on $\mathbf{c}^n$ is given by $< u, v > = u^t \bar{v}$. So, we calculate:

$$\begin{aligned} \bar{\lambda} < v, v > & = v^t \lambda \bar{v} = v^t \overline{(Sv)} \\ & = v^t S \bar{v} = \left(S^t v\right)^t \bar{v} \\ & = (Sv)^t \bar{v} = \lambda v^t \bar{v} \\ & = \lambda < v, v >. \end{aligned} \tag{6.50}$$

Now, since $< v, v > \neq 0$ we see $\lambda = \bar{\lambda}$ so λ is real and hence, we can choose $v \in \mathbf{R}^n$.

Using this lemma, we can decompose $\mathbf{R}^n$ into invariant subspaces of $A \in SO(n)$.

LEMMA 6.24   *Let $A \in SO(n)$. Then there is $W \subset \mathbf{R}^n$ which is an invariant subspace of A with* $\dim W = 1$ *or* 2.

Proof   If A has an eigenvector v let $W = \mathbf{R}v$ and we are finished. If A does not have an eigenvector, set

$$S = A + A^t. \tag{6.51}$$

Then S is a real symmetric matrix and so, by Lemma 6.23, S has an eigenvector v. Set

$$W = Sp(v, Av). \tag{6.52}$$

Then, $\dim W = 2$ and it only remains to show that W is invariant under A. Clearly, for $v \in W$, $Av \in W$, so it is only necessary to show $A^2v \in W$. Since $A \in SO(n)$ $A^t = A^{-1}$, and therefore,

$$Av + A^{-1}v = \lambda v. \tag{6.53}$$

Hence,

$$A^2v = \lambda Av - v, \tag{6.54}$$

and therefore, $A^2v \in W$ and the proof is complete.

This lemma by induction gives the following one.

LEMMA 6.25   *Let $A \in SO(n)$. Then, A is conjugate to a matrix of the form*

$$\begin{pmatrix} 1 & & & & & & & & \\ & \ddots & & & & & & & \\ & & 1 & & & & & & \\ & & & -1 & & & & & \\ & & & & \ddots & & & & \\ & & & & & -1 & & & \\ & & & & & & \cos\theta_1 & -\sin\theta_1 & \\ & & & & & & \sin\theta_1 & \cos\theta_1 & \\ & & & & & & & & \ddots \end{pmatrix}.$$

Proof   It is sufficient to find mutually orthogonal spaces $W_1, W_2, \ldots, W_r$ such that each W_i is invariant under A and $\dim W_i = 1$ or 2. If $\dim W_i = 1$ then $W_i = \mathbf{R}v$ and v is an eigenvector of A. Since $A \in$

$SO(n)$ its real eigenvalues are either 1 or -1. Label the W_i so that the spaces corresponding to eigenvalues 1 come first, then the spaces corresponding to eigenvalue -1, and finally the two-dimensional spaces. If W is an invariant subspace of $\mathbb{R}^n$ for A then, since $A \in SO(n)$, $W^\perp$ is also an invariant subspace. Thus, using Lemma 6.24, we obtain one space W and by induction $A \mid W^\perp$ has the form of the result in this lemma.

Remark 6.26   There is an even number of -1 in the matrix in Lemma 6.25. This follows by taking determinants. If there are j entries of -1 then

$$\det \begin{pmatrix} 1 & & & & & & & \\ & \ddots & & & & & & \\ & & 1 & & & & & \\ & & & -1 & & & & \\ & & & & \ddots & & & \\ & & & & & -1 & & \\ & & & & & & \cos\theta_1 & -\sin\theta_1 \\ & & & & & & \sin\theta_1 & \cos\theta_1 \\ & & & & & & & & \ddots \end{pmatrix} = (-1)^j. \quad (6.55)$$

Thus, $1 = \det A = (-1)^j$ and so j is even. We can write the 2×2 blocks

$$\begin{pmatrix} -1 & 0 \\ 0 & -1 \end{pmatrix} = \begin{pmatrix} \cos\theta & -\sin\theta \\ \sin\theta & \cos\theta \end{pmatrix} \text{ with } \theta = \pi. \quad (6.56)$$

Thus, if $A \in SO(n)$ we can write

$$A = P \begin{pmatrix} 1 & & & & \\ & \ddots & & & \\ & & 1 & & \\ & & & \cos\theta & -\sin\theta \\ & & & \sin\theta & \cos\theta \\ & & & & & \ddots \end{pmatrix} P^{-1}, \quad (6.57)$$

where some of the θ may be π. If we define

$$A_t = P \begin{pmatrix} 1 & & & & \\ & \ddots & & & \\ & & 1 & & \\ & & & \cos t\theta & -\sin t\theta \\ & & & \sin t\theta & \cos t\theta \\ & & & & & \ddots \end{pmatrix} P^{-1}, \quad (6.58)$$

we see that the map $\gamma : [0,1] \to SO(n)$ by $\gamma(t) = A_t$ is a path joining the identity matrix $\gamma(0) = I$ to $\gamma(1) = A$. Thus, we have shown $SO(n)$ is connected.

To see $O(n)$ has two connected components, let

$$D = \begin{pmatrix} -1 & & & \\ & 1 & & \\ & & \ddots & \\ & & & 1 \end{pmatrix}.$$ (6.59)

Then $\det D = -1$. If $\gamma : [0,1] \to O(n)$ is a path with $\gamma(0) = I$ and $\gamma(1) = D$ then $\det(\gamma(t))$ is a continuous map such that $\det \gamma(0) = 1$, $\det \gamma(1) = -1$ and $\det \gamma(t) = +1$ or -1 for all t. This is not possible and we see that $O(n)$ is not connected. If $A \in O(n)$ then either $A \in SO(n)$ (the case $\det A = 1$) or $DA \in SO(n)$ (the case $\det A = -1$). Thus, if $A \in O(n)$ then $A \in SO(n)$ or $A = DB$ with $B \in SO(n)$. That is, we have shown

$$O(n) = SO(n) \cup DSO(n)$$ (6.60)

where $DSO(n) = \{DB : B \in SO(n)\}$. In other words, $O(n)$ is the union of two connected components.

Example 6.27   The group $Sp(n)$ is connected.

To see this, we start with $Sp(1)$. Define $\psi : Sp(1) \to SU(2)$ by

$$\psi(a + ib + jc + kd) = \begin{pmatrix} a + ib & c + id \\ -c + id & a - ib \end{pmatrix}.$$ (6.61)

Then, ψ is the inverse to the map ϕ given after equation (2.28) and is a group isomorphism. Thus, we see that $Sp(1)$ is connected. However, before moving on to the case of $Sp(n)$ we need to make some observations on the map ψ.

Lemma 6.28   *If $q \in \mathbb{H}$ is a quaternion of unit length, there is a quaternion $\tilde{q}$ of unit length so that $\tilde{q}q\tilde{q}^{-1}$ is a complex number of unit length.*

Proof   Since $q \in Sp(1)$ then $\psi(q) \in SU(2)$. Thus

$$P\,\psi(q)P^{-1} = \begin{pmatrix} e^{i\theta} & 0 \\ 0 & e^{-i\theta} \end{pmatrix}.$$ (6.62)

Let $\tilde{q} = \psi^{-1}(P)$. Then (6.62) becomes

$$\psi(\tilde{q})\psi(q)\psi\left(\tilde{q}^{-1}\right) = \psi\left(e^{i\theta}\right),$$ (6.63)

which completes the proof since ψ is a group isomorphism.

Next, we define $\sigma : \mathsf{H} \to \mathsf{C}^2$ by

$$\sigma(a + ib + jc + kd) = \begin{pmatrix} a + ib \\ -c + id \end{pmatrix}. \qquad (6.64)$$

LEMMA 6.29   *If $q \in Sp(1)$ and $v \in \mathsf{H}$ then $\sigma(qv) = \psi(q)\sigma(v)$.*

Proof   This is a straightforward computation analogous to showing that ψ is a group homomorphism.

For the case of $Sp(n)$ we extend both ψ and σ to $\psi : Sp(n) \to SU(2n)$ and $\sigma : \mathsf{H}^n \to \mathsf{C}^{2n}$. Since in practice no confusion arises, the same symbol is used for the extension as for the original map.

LEMMA 6.30   *The map $\psi : Sp(n) \to SU(2n)$ is a group homomorphism (but not an isomorphism if $n \neq 1$) and $\sigma(Av) = \psi(A)\sigma(v)$ for $A \in Sp(n)$ and $v \in \mathsf{H}^n$.*

Proof   This is again a straightforward calculation and is left as an exercise.

LEMMA 6.31   *Let $A \in Sp(n)$. Then, there is $v \in \mathsf{H}^n$ so $v\mathsf{H}$ is invariant under A.*

Proof   Let u be an eigenvector for $\psi(A)$ so,

$$\psi(A)u = \lambda u \qquad (6.65)$$

with λ a unit length complex number. Set $v = \sigma^{-1}(u)$.

Let $S \in \mathsf{H}$ and set $r = |s|$ and $q = s/r$ so $q \in Sp(1)$. Then, we calculate

$$\begin{aligned} \sigma(Avs) &= \sigma(Avr(qI)) \\ &= \psi(A)ur\psi(qI) \\ &= \lambda ur\psi(qI) \\ &= \sigma(vrq\lambda). \end{aligned} \qquad (6.66)$$

The last step follows from a calculation which shows $\sigma(v\lambda) = \sigma(v)\lambda$. See Exercise 6.10.

THEOREM 6.32   *Let $A \in Sp(n)$. Then, A is conjugate to a matrix*

$$\begin{pmatrix} \lambda_1 & & \\ & \ddots & \\ & & \lambda_n \end{pmatrix}$$

where λ_i is a unit length complex number.

Proof   By induction using Lemma 6.31, we can find a linearly independent set of vectors $v_1, \ldots, v_n \in \mathsf{H}^n$ so that for $s_i \in \mathsf{H}$ we have

$$A\left(\sum v_i s_i\right) = \sum v_i s_i \lambda_i. \tag{6.67}$$

Since λ_i is an eigenvalue of $\psi(A) \in SU(n)$ it is a unit length complex number. Furthermore, as in the case of $U(n)$, we can choose $v_i, \ldots, v_n$ to be an orthonormal set of vectors.

Let

$$P = (v_1,\ v_2, \ldots, v_n) \tag{6.68}$$

be the matrix with the vectors $v_1, \ldots, v_n$ as columns. Then $P \in Sp(n)$ and equation (6.67) gives

$$P^{-1}AP = \begin{pmatrix} \lambda_1 & & \\ & \ddots & \\ & & \lambda_n \end{pmatrix}, \tag{6.69}$$

which completes the proof.

Corollary 6.33   *The group $Sp(n)$ is connected.*

Proof   With AP and $\lambda_1, \ldots, \lambda_n$ as in Theorem 6.32, set

$$A_t = P \begin{pmatrix} \lambda_1^t & & \\ & \ddots & \\ & & \lambda_n^t \end{pmatrix} P^{-1}. \tag{6.70}$$

Then $A_0 = I$, $A_1 = A$ and $A_t \in Sp(n)$ for all $t \in [0,1]$. Thus $t \to A_t$ is a path joining A to the identity in $Sp(n)$.

Exercises - Chapter 6

6.1   Let Z be the center of G, a compact connected Lie group, and T a maximal torus of G. Show that $Z \subset T$.

6.2   Continue Exercise 6.1 and show that $Z = \cap xTx^{-1}$ where the intersection is over all $x \in G$.

6.3   Use Exercises 5.6 and 6.1 to show that the center of $U(n)$ is $S^1 = \{e^{i\theta}I\}$.

6.4   Show that the center of $SU(n)$ is $\mathsf{z}_n = \{e^{2\pi i\theta}I : n\theta \in \mathsf{z}\}$.

6.5   Show directly that $\exp : \mathcal{U}(n) \to U(n)$ is onto by using the fact that any element of $U(n)$ is conjugate to an element in the maximal torus and $\exp : T \to T$ is onto.

6.6   Let $A \in U(n)$. Show that $\mathbb{C}^n$ has an Hermitian orthonormal basis of eigenvectors of A. Hence, show that A is conjugate to an element in the maximal torus and so $U(n)$ is connected.

6.7   Show $SU(n)$ is connected.

6.8   Recall from Exercise 2.10 that $\begin{pmatrix} z & w \\ -\bar{w} & \bar{z} \end{pmatrix} \in SU(2)$ is conjugate to $\begin{pmatrix} e^{i\theta} & 0 \\ 0 & e^{-i\theta} \end{pmatrix}$, where $2\cos\theta = z+\bar{z}$. Now, find explicitly $\tilde{q}$ of Lemma 6.28 so $\tilde{q}q\tilde{q}^{-1}$ is a complex number of unit length.

6.9   Give the details of the proof of Lemma 6.30. That is, prove $\psi(AB) = \psi(A)\psi(B)$ and $\sigma(Av) = \psi(A)\sigma(v)$.

6.10   Show for $\lambda \in \mathbb{C}$ and $v \in \mathbb{H}$ that $\sigma(v\lambda) = \sigma(v)\lambda$. Hence, show the same result holds for $v \in \mathbb{H}^n$.

Chapter 7

Representation Theory

This chapter begins with a long list of definitions followed by some elementary results about representation theory, which lead to the Schur orthogonality relations. While these results are elementary, they are, in fact, very useful. Our point of view has been to develop the theory of compact Lie groups, so that one can carry out analysis on them. A major tool for this is the Peter- Weyl theorem, which is discussed in this chapter. Representations of the torus are described at the end of this chapter.

The definitions are as follows:

DEFINITION 7.1   A finite dimensional representation is a continuous homomorphism $\pi : G \to Aut\, V$, where V is a vector space.

DEFINITION 7.2   The character of a representation is $\chi(x) = tr\, \pi(x)$, so χ is a map from G to the field of scalars of V.

DEFINITION 7.3   A subspace V_0 is invariant if $\pi(x)V_0 \subset V_0$ for all $x \in G$. There are always at least two invariant subspaces of any representation: $V_0 = V$ and $V_0 = \{0\}$. These are the trivial invariant subspaces.

DEFINITION 7.4   A representation is irreducible if it has no non-trivial invariant subspaces.

DEFINITION 7.5   An innerproduct $< , >$ on V is invariant if $< \pi(x)u, \pi(x)v > = < u, v >$ for all $x \in G$ and $u, v \in V$.

THEOREM 7.6   *If $\pi : G \to Aut\, V$ is a representation, then V has an invariant innerproduct.*

Proof Let $(u \mid v)$ be any innerproduct on V. Then, define $< u, v >$ by

$$< u, v > = \int_G (\pi(x)u \mid \pi(x)v)\, dx. \qquad (7.1)$$

With this definition $< u, v >$ is an invariant innerproduct. Here, we have taken $\int_G dx$ to be the integral defined by the Killing form metric and, for convenience, normalized, so that $\int_G 1 dx = 1$. Notice that this proof requires that G be compact, so that we can use a normalized integral.

A consequence of this is the following result. Note that this result fails to be true for infinite dimensional representations.

THEOREM 7.7 *If π is a finite dimensional representation, then π is a direct sum of irreducible representations.*

Proof Let $\pi : G \to Aut\, V$. Then, we proceed by induction on $\dim V$. Clearly, if $\dim V = 1$ the result is trivially true. If V_0 is a non-trivial invariant subspace, let $V_1 = (V_0)^\perp$, using the invariant innerproduct. Now V_1 is an invariant subspace since

$$< \pi(g)v, u > = < v, \pi\left(g^{-1}\right)u > = 0 \qquad (7.2)$$

for $u \in V_0$, $v \in V_1$ and $g \in G$. By our inductive assumption, both V_0 and V_1 can be written as a sum of irreducible representations. Hence V can be written as a sum of irreducible representations.

So far, all of these results hold for either complex or real representations. This will be true for most of our results. However, there are some results which only hold for complex representations. When we come to these, we shall point out that the representations are complex. We can now state and prove the Schur orthogonality relations.

DEFINITION 7.8 The representation $\pi_1 : G \to Aut\, V_1$ is equivalent to $\pi_2 : G \to Aut\, V_2$ if there is an invertible $P \in \text{Hom}\,(V_1, V_2)$ such that $\pi_1(g) = P^{-1}\pi_2(g)P$ for all $g \in G$. If two representations are not equivalent, they are inequivalent.

DEFINITION 7.9 The space $\text{Hom}_G\,(V_1, V_2)$ is the set of $f \in \text{Hom}\,(V_1, V_2)$ such that $f\left(\pi_1(g)v\right) = \pi_2(g)f(v)$.

THEOREM 7.10 *(Schur orthogonality relations). If $\pi_1 : G \to Aut\, V_1$ and $\pi_2 : G \to Aut\, V_2$ are inequivalent irreducible representations, then*

$\mathrm{Hom}_G(V_1, V_2) = 0.$

Proof Let $f \in \mathrm{Hom}_G(V_1, V_2)$ so that $f(\pi_1(x)v) = \pi_2(x)f(v)$. Then, $\mathrm{Ker} f$ is an invariant subspace of V_1 so, by irreducibility $\mathrm{Ker} f = V_1$ or $\mathrm{Ker} f = 0$. If $\mathrm{Ker} f = V_1$ then $f = 0$ and the proof is complete. If $\mathrm{Ker} f = 0$ then consider $\mathit{Im}\ f$. This is an invariant subspace of V_2. Thus, by irreducibility either $\mathit{Im}\ f = V_2$ or $\mathit{Im}\ f = 0$. However, $\mathit{Im}\ f = V_2$ is impossible, since f is then invertible and V_1 is not equivalent to V_2. If $\mathit{Im}\ f = 0$ then $V_1 = 0$ and the result is trivially true.

COROLLARY 7.11 *Let V_i be the irreducible representations of G and $\oplus n_i V_i$ be equivalent to $\oplus m_i V_i$ then $n_i = m_i$.*

Proof Since $\oplus n_i V_i$ is equivalent to $\oplus m_i V_i$ we have

$$\mathrm{Hom}_G(\oplus n_i V_i, V_j) \cong \mathrm{Hom}_G(\oplus m_i V_i, V_j). \qquad (7.3)$$

Thus

$$\oplus n_i \mathrm{Hom}_G(V_i, V_j) \cong \oplus m_i \mathrm{Hom}_G(V_i, V_j). \qquad (7.4)$$

The result of Corollary (7.11) now follows by taking dimensions in equation (7.4).

In the remaining results of this chapter, the complex numbers play a special role. As we remarked before, some of the results are only true over the complex numbers. Others, for example, the next result, are true in both the real and complex cases. However, the complex case is more easily proved. Then, the real case can often be deduced from the complex one. The next result is the Schur orthogonality relations for characters.

THEOREM 7.12 *Let $\pi_v : G \to Aut\ V$ and $\pi_w : G \to Aut\ W$ be two representations with characters χ_v and χ_w. Then,*

$$\int_G \overline{\chi_v}(g)\chi_w(g)dg = \dim\ \mathrm{Hom}_G(V, W),$$

where the bar denotes complex conjugation in the complex case and nothing in the real case.

Proof First, we prove the result in the complex case. Notice that $\overline{\chi_V(g)} = \chi_{V^*}(g)$ where V^* is the dual representation. Then $\overline{\chi_V}(g)\chi_W(g) = \chi_{V^* \otimes W}(g)$. Now, the result reduces to

$$\int_G \chi_{V^* \otimes W}(g)dg = \dim(V^* \otimes W)^G. \qquad (7.5)$$

Here, the superscript G denotes the G-fixed vectors. We shall prove that for any representation $\pi : G \to Aut\, V$ we have

$$\int_G \chi(g)dg = \dim V^G. \tag{7.6}$$

To prove (7.6) define $T : V \to V$ by

$$T(v) = \int_G \pi(g)v\, dg. \tag{7.7}$$

Now T is idempotent and $Im\, T = V^G$. However,

$$\int_G \chi(g)dg = \int_G tr\, \pi(g)dg \tag{7.8}$$
$$= tr\, T. \tag{7.9}$$

Therefore, (7.6) is true. The real case follows by complexifying, that is, taking tensor products with c over R. The key step is to notice that

$$\dim \mathrm{Hom}_{\mathbf{c},G}\left(V \otimes_{\mathbf{R}} \mathbf{c}, W \otimes_{\mathbf{R}} \mathbf{c}\right) = \dim \mathrm{Hom}_{\mathbf{R},G}(V, W). \tag{7.10}$$

Remark 7.13   It follows from the Schur orthogonality relations that the functions χ_V for inequivalent irreducible V are linearly independent.

Definition 7.14   A function f on G is a class function if $f\left(gxg^{-1}\right) = f(x)$.

Since for any linear transformations $tr(AB) = tr(BA)$, it is now easy to see that characters are class functions. The next result is very useful. It says, essentially, that the set of all characters acts as a basis for the space of all continuous class functions on G. This is a generalization of the results on Fourier series for periodic functions of a single variable.

Theorem 7.15   *(Peter–Weyl Theorem).   Any continuous class function $f : G \to \mathbf{c}$ can be uniformly approximated by characters.*

Proof   We omit the proof of this result and refer the reader to a more advanced work..

The next two results are only true over the complex numbers. These results then enable us to fully describe all the irreducible representations of the torus.

THEOREM 7.16 *Let V be an irreducible complex representation space of G. If $f \in \mathrm{Hom}_G(V, V)$ then $f(V) = cV$ for some $c \in \mathbf{c}$.*

Proof Consider $f : V \to V$. Then, by the fundamental theorem of algebra f has an eigenvalue. Let this eigenvalue be $c \in \mathbf{c}$. Then $f - cI$ is a singular invariant homomorphism of an irreducible representation. Therefore, by the same argument as in the proof of Theorem 7.10, $f - cI = 0$.

COROLLARY 7.17 *If G is abelian and π is an irreducible complex representation, then π is one-dimensional.*

Proof For any $g \in G$ the homomorphism

$$\pi(g) : V \to V \tag{7.11}$$

is invariant. This is because G is abelian. Thus $\pi(g)$ is a multiplication by a scalar $c(g)$. This means that every subspace of V is invariant, and therefore, by irreducibility V is one-dimensional.

To start our work on the representations of the torus, we begin with the simplest case, that of the circle.

LEMMA 7.18 *Let $\alpha : S^1 \to S^1$ be a continuous homomorphism. Then, $\alpha(x) = nx$ for some $n \in \mathbf{z}$.*

Note: Here we are identifying

$$S^1 = \mathbf{R}/\mathbf{z}. \tag{7.12}$$

Proof Lift α to $\beta : \mathbf{R} \to \mathbf{R}$. Then $\beta(1) = 0 \bmod \mathbf{z}$. Let $n = \beta(1)$. Then $\beta(a) = n(a)$ for $a \in \mathbf{z}$. Since $\beta(1/q) = \beta(1)/q$ for $q \in \mathbf{z}$ we have $\beta(p/q) = np/q$ for any rational p/q. The result now follows by continuity.

COROLLARY 7.19 *If $\alpha : T \to S^1$ is a continuous homomorphism, then $\alpha(x_1, \ldots, x_k) = n_1 x_1 + \ldots + n_k x_k$.*

Proof This is a routine application of the relation $\alpha(x + y) = \alpha(x) + \alpha(y)$.

THEOREM 7.20 *If π is a complex irreducible representation of T. Then,*

$$\pi(x_1, \cdots, x_k) = \exp\left(2\pi i \left(n_1 x_1 + \cdots + n_k x_k\right)\right).$$

Proof   So far, we have identified $S^1 = \mathbf{R}/\mathbf{Z}$. However, S^1 acts on C naturally as multiplication by a unit complex number. This corresponds to the identification

$$S^1 = \{\exp(2\pi i t) : t \in \mathbf{R}/\mathbf{Z}\}. \tag{7.13}$$

Since π is one-dimensional (Corollary 7.17) π factors as $\pi = m\alpha$ where $\alpha : T \to S^1$ is a homomorphism and m is a multiplication using the identification (7.13) .

The next result is the analogous result in the real case.

Theorem 7.21   *Let π be a real irreducible representation of T. Then π is either trivial or is two-dimensional. In the case where π is two-dimensional, it is just a complex representation viewed as a real one.*

Proof   First, note that the irreducible complex representations, when viewed as real ones, are still irreducible. This fact is not true for a general Lie group, but is true in this special case of a torus. Let V be an irreducible real representation. Then

$$V_{\mathbf{C}} = V \otimes_{\mathbf{R}} \mathbf{C} = \oplus \, n_i \alpha_i. \tag{7.14}$$

Now

$$\dim_{\mathbf{C}} \mathrm{Hom}_{\mathbf{C},G} \left(V_{\mathbf{C}}, \alpha_i \right) = \dim_{\mathbf{R}} \mathrm{Hom}_{\mathbf{R},G} \left(V, \alpha_i \right). \tag{7.15}$$

Thus, either $V = \alpha_i$ or $n_i = 0$ by the Schur orthogonality relations.

We complete this chapter with the following example:

Example 7.22   The representations of $SU(2)$.

We shall describe all the complex irreducible representations of $SU(2)$. First, we note that $SU(2)$ has a natural action on $\mathbf{c}^2$:

$$\begin{pmatrix} z & w \\ -\bar{w} & \bar{z} \end{pmatrix} \begin{pmatrix} x \\ y \end{pmatrix} = \begin{pmatrix} zx & + & wy \\ -\bar{w}x & + & \bar{z}y \end{pmatrix}. \tag{7.16}$$

Let $\mathcal{F}$ denote the set of complex valued functions on $\mathbf{c}^2$. Then (7.16) induces a representation π of $SU(2)$ on $\mathcal{F}$:

$$\pi \begin{pmatrix} z & w \\ -\bar{w} & \bar{z} \end{pmatrix} f \begin{pmatrix} x \\ y \end{pmatrix} = f \left(\begin{pmatrix} \bar{z} & -w \\ \bar{w} & z \end{pmatrix} \begin{pmatrix} x \\ y \end{pmatrix} \right). \tag{7.17}$$

This is just the action $\pi(A)f = f \circ A^{-1}$ and is easily seen to be a representation. Let $V_n \subset \mathcal{F}$ be the subspace of homogeneous polynomials of degree n. Then, a basis for V_n is

$$x^n, x^{n-1}y, x^{n-2}y^2, \cdots, x^{n-r}y^r, \cdots, y^n \tag{7.18}$$

and

$$\dim V_n = n + 1. \tag{7.19}$$

Clearly V_n is an invariant subspace of $\mathcal{F}$ for the representation π. Let π_n denote the restriction of π to V_n. Then

$$\pi_n \begin{pmatrix} z & w \\ -\bar{w} & \bar{z} \end{pmatrix} \sum c_r x^{n-r} y^r = \sum c_r (\bar{z}x - wy)^{n-r} (\bar{w}x + zy)^r. \tag{7.20}$$

Let $\chi_n(g) = tr \; \pi_n(g)$ be the character of π_n. Then, restricted to the maximal torus, we find:

$$\chi_n \begin{pmatrix} e^{i\theta} & 0 \\ 0 & e^{-i\theta} \end{pmatrix} = \sum_{r=0}^{n} e^{i(n-2r)\theta} = \frac{e^{i(n+1)\theta} - e^{-i(n+1)\theta}}{e^{i\theta} - e^{-i\theta}}. \tag{7.21}$$

In fact, the representations π_n are all the irreducible representations of $SU(2)$. The easiest way to see this is to use the highest weight theorem of Chapter 8, and the Weyl formulae of Chapter 9. Using these, we shall see that $SU(2)$ has irreducible representations for each non-negative integer n, that the irreducible representation indexed by n has dimension $n+1$ and that the character of this irreducible representation is given by the formula (7.21). Together these facts establish that we have described all the complex irreducible representations of $SU(2)$.

Exercises - Chapter 7

7.1 Let $\pi : SU(2) \rightarrow GL(4, \mathbf{c})$ by $\pi \begin{pmatrix} z & w \\ -\bar{w} & \bar{z} \end{pmatrix}$

$$= \begin{pmatrix} z^2 & zw & zw & w^2 \\ -z\bar{w} & z\bar{z} & -w\bar{w} & w\bar{z} \\ -z\bar{w} & -w\bar{w} & z\bar{z} & w\bar{z} \\ \bar{w}^2 & -w\bar{z} & -w\bar{z} & \bar{z}^2 \end{pmatrix}.$$ Show that π is a representation of $SU(2)$.

7.2 Let $V_1 = \left\{ \begin{pmatrix} 0 \\ a \\ -a \\ 0 \end{pmatrix} : a \in \mathbf{c} \right\}$ and $V_2 = \left\{ \begin{pmatrix} b \\ c \\ c \\ d \end{pmatrix} : b, c, d \in \mathbf{c} \right\}$. Show that V_1 and V_2 are invariant subspaces of $\mathbf{c}^4$ for the representation π

of Exercise 7.1.

7.3   Show that V_1 and V_2 of Exercise 7.2 are irreducible.

7.4   Find the trace of $\pi_1 = \pi \mid V_1$ and $\pi_2 = \pi \mid V_2$.

7.5   Let χ_n be the character of π_n for $SU(2)$. Use Exercise 2.10 to calculate $\chi_n \begin{pmatrix} z & w \\ -\bar{w} & \bar{z} \end{pmatrix}$.

7.6   Let V_m and V_n be two representation spaces for $SU(2)$ as defined after equation (7.17) with $m \neq n$. Let $T : V_m \to V_n$ be an $SU(2)$ invariant linear transformation. Show by a direct computation (without using the Schur orthogonality relations) that $T = 0$.

7.7   Let $\pi_m : SU(2) \to V_m$ and $\pi_n : SU(2) \to V_n$ be two representations of $SU(2)$. Then, form the tensor product representation $\pi_m \otimes \pi_n : SU(2) \to V_m \otimes V_n$ and show $tr\,(\pi_m \otimes \pi_n(g)) = tr\,\pi_m(g) tr\,\pi_n(g)$.

7.8   From Exercise 7.5 we have $\chi_n \begin{pmatrix} z & w \\ -\bar{w} & \bar{z} \end{pmatrix} = \dfrac{\sin(n+1)\theta}{\sin\theta}$, where $2\cos\theta = z + \bar{z}$. Show that for $n \geq m$, $\chi_m(A)\chi_n(A) = \chi_{n-m}(A) + \chi_{n-m+2}(A) + \ldots + \chi_{n+m}(A)$. (Hint: $\sin\alpha + \sin(\alpha + \beta) + \sin(\alpha + 2\beta) + \ldots + \sin(\alpha + k\beta) = \sin(\alpha + \frac{1}{2}\,k\beta)\,\sin(\frac{1}{2}(k+1)\beta)\,/\sin(\frac{1}{2}\,\beta).$)

7.9   From Exercise 7.8 show that for $SU(2)$ and $n \geq m$ that $\pi_m \otimes \pi_n = \pi_{n-m} \oplus \pi_{n-m+2} \oplus \ldots \oplus \pi_{n+m}$.

7.10   Show that the representation of Exercise 7.1 is $\pi_1 \otimes \pi_1$.

Chapter 8

Roots and Weights

In this chapter, we introduce the weights of a representation. When the representation is the adjoint representation, the weights are called roots. After introducing these, we then define several lattices, the integer lattice $I \subset T$, the lattice of weights $P \subset T^*$ and the lattice of roots $Q \subset T^*$. It happens that there is a partial ordering on the weights of a representation such that each irreducible representation has a highest weight. This highest weight is uniquely determined by the representation and, in turn, determines the representation. Since there is a relatively simple description of the set of highest weights in terms of P, we have a complete description of the irreducible representations of G.

Let $\pi : G \to Aut\ V$ be a representation and let $\pi \mid T$ denote the restriction of π to a maximal torus T. This restriction decomposes into irreducible representations of T as follows:

$$\pi(x) = \begin{pmatrix} 1 & & & & \\ & 1 & & & \\ & & \Theta_1(x) & & \\ & & & \Theta_k(x) \end{pmatrix}, \tag{8.1}$$

where $\Theta_j(x)$ denotes the 2×2 block

$$\Theta_j(x) = \begin{pmatrix} \cos\theta_j(x) & -\sin\theta_j(x) \\ \sin\theta_j(x) & \cos\theta_j(x) \end{pmatrix}. \tag{8.2}$$

In this 2×2 block θ_j is a homomorphism

$$\theta_j : T \to \text{R}/2\pi\text{Z} \tag{8.3}$$

which we differentiate to give

$$d\theta_j : \mathcal{T} \to \text{R}. \tag{8.4}$$

69

DEFINITION 8.1   The set $\{\pm d\theta_j\}$ is the set of weights of the representation π.

Notation 8.2   *Sometimes, we shall refer to $\{\pm\theta_j\}$ as the weights of π and, usually, we shall suppress the d to write θ_j in place of $d\theta_j$.*

This notational convention is, of course, confusing and ambiguous. However, in practice it causes no trouble to use the same symbol, θ_j, for both θ_j and $d\theta_j$. Equally well, no difficulty is caused by calling either θ_j or $d\theta_j$ a weight of π. We can distinguish between these by looking at the domain of the weight. If we have θ_j when we need $d\theta_j$ all we have to do is differentiate θ_j. Conversely, θ_j can be found from $d\theta_j$ by integration, locally, and using the fact that θ_j is a homomorphism.

DEFINITION 8.3   If $\pi = Ad$, the adjoint representation, then the weights of π are called roots and are usually denoted by $\{\pm\alpha_i\}$.

We now introduce some lattices in the Lie algebra $\mathcal{T}$ of T and its dual $\mathcal{T}^*$.

DEFINITION 8.4   The integer lattice $I \subset \mathcal{T}$ is given by $2\pi I = \exp^{-1}(1)$.

DEFINITION 8.5   In $\mathcal{T}^*$ the lattice P is $P = \{f \in \mathcal{T}^* : f(I) \subset \mathbf{z}\}$.

LEMMA 8.6   *If θ is the weight of a representation, then $\theta \in P$.*

Proof   Let $x \in I$. Then,

$$\pi(\exp 2\pi x) = \begin{pmatrix} 1 & & \\ & \ddots & \\ & & 1 \end{pmatrix}, \tag{8.5}$$

for any representation π. Thus, if θ is a weight of π, then

$$2\pi\theta(x) = \theta(\exp 2\pi x) = 0 \bmod 2\pi\mathbf{z}, \tag{8.6}$$

where the second $\theta : T \to \mathbf{R}/2\pi\mathbf{z}$. Hence $2\pi\theta(x) = 2\pi n$ so $\theta(x) = n$.

DEFINITION 8.7   The lattice $Q \subset \mathcal{T}^*$ is the lattice of roots, $Q = \oplus \mathbf{z}\alpha_i$ over all roots α_i.

Clearly, $Q \subset P$. It is also true, but not obvious, that P is, in fact, the lattice generated by the weights of all representations. This statement is, essentially, a converse to Lemma 8.6, and will be stated more precisely in the highest weight theorem at the end of this chapter.

We now introduce some hyperplanes in T. If $\alpha \in T^*$ let

$$L_\alpha = \{x \in T : \alpha(x) = 0\}. \tag{8.7}$$

We are particularly interested in L_α when α is a root.

LEMMA 8.8 *The space $T - \cup L_\alpha$, with the union over all roots α, is a disjoint union of cones.*

Proof This is clear, since L_α is a hyperplane in T.

DEFINITION 8.9 The Weyl group $W = N(T)/T$, where $N(T)$ is the normalizer of T in G.

The normalizer, $N(T)$, is the largest subgroup of G to act on T by conjugation. Clearly, since a torus is abelian, T acts trivially on itself by conjugation. Thus, W acts on T, and hence, on T.

DEFINITION 8.10 The connected components of $T - \cup L_\alpha$ are called Weyl chambers. Notice that these are open sets in T.

THEOREM 8.11 *The Weyl group W permutes the Weyl chambers.*

Proof Let $w \in W$ and let $n \in N(T)$ be a representative of w. Then, w acts on T in the same way as $Ad\, n$. Suppose our decomposition for $Ad \mid T$ is

$$Ad\, g = \begin{pmatrix} 1 & & & & & \\ & \ddots & & & & \\ & & 1 & & & \\ & & & \alpha_1(g) & & \\ & & & & \ddots & \\ & & & & & \alpha_k(g) \end{pmatrix}, \tag{8.8}$$

where there are ℓ entries which are identically 1 and $\alpha_j(g)$ represents the 2×2 block given by replacing θ_j by α_j in (8.2). Then, for x sufficiently close to $0 \in T$,

$$\dim \ker(Ad(\exp x) - I) = \ell + 2r, \tag{8.9}$$

where I is the identity matrix and r is the number of hyperplanes which contain x. Now, we calculate:

$$\dim \ker(Ad(\exp wx) - I)$$
$$= \dim \ker(Ad(\exp(Adn)x) - I)$$
$$= \dim \ker \left(Ad\left(n(\exp x)n^{-1}\right) - I\right)$$
$$= \dim \ker \left(Adn(Ad \exp x - I)Adn^{-1}\right)$$
$$= \dim \ker(Ad(\exp x) - I). \tag{8.10}$$

From this calculation, we see that x and wx are in the same number of hyperplanes. In particular, this proves that W permutes the Weyl chambers.

The number ℓ of this proof is the dimension of the maximal torus. It is called the rank of G.

In fact, the connection between the Weyl group and Weyl chambers is even stronger than was proved in the previous theorem. The Weyl group acts simply transitively on the Weyl chambers. Thus, there is a one-to-one correspondence between elements of the Weyl group and the Weyl chambers. However, this correspondence is not canonical. The action of the Weyl group is given by reflections in the hypersurfaces, which are called the walls of the Weyl chambers. To show the content of these remarks more clearly, we shall prove that the reflections in the hyperplanes L_α are in the Weyl group and from this deduce that the Weyl group acts transitively on the Weyl chambers. The fact that the Weyl group is generated by a subset of these reflections and that it acts simply transitively on the Weyl chambers shall be left to more advanced texts for their proofs.

THEOREM 8.12   *Let $r_\alpha : \mathcal{T} \to \mathcal{T}$ be a reflection in the hyperplane L_α then $r_\alpha \in W$.*

Proof   Fix a root α and then make the following definitions. Let $v_\alpha \in \mathcal{T}$ which is perpendicular to L_α such that

$$\alpha(x) = \langle x, v_\alpha \rangle \tag{8.11}$$

for all $x \in \mathcal{T}$. In the decomposition given by (8.1), let V_α be the two-dimensional subspace of $\mathcal{G}$ which is the representation space associated to the root α. Pick a basis $\{X_\alpha, Y_\alpha\}$ of V_α such that the adjoint representation has matrix

$$Ad\, g|V_\alpha = \begin{pmatrix} \cos \alpha(g) & -\sin \alpha(g) \\ \sin \alpha(g) & \cos \alpha(g) \end{pmatrix} \tag{8.12}$$

for $g \in T$ when $Ad\, g$ is restricted to V_α. Then, for $x \in T$ we see the matrix of $ad\, x$ restricted to V_α is

$$ad\, x | V_\alpha = \begin{pmatrix} 0 & -\alpha(x) \\ \alpha(x) & 0 \end{pmatrix}. \tag{8.13}$$

That is

$$\begin{aligned} [x, X_\alpha] &= \alpha(x) Y_\alpha, \\ [x, Y_\alpha] &= -\alpha(x) X_\alpha. \end{aligned} \tag{8.14}$$

Now, by the Jacobi identity

$$[x, [X_\alpha, Y_\alpha]] = -[Y_\alpha, [x, X_\alpha]] - [X_\alpha, [Y_\alpha, x]] = 0. \tag{8.15}$$

Thus, we have $[X_\alpha, Y_\alpha] \in T$. Now, using the invariance of the Killing form innerproduct, we have:

$$\begin{aligned} \langle x, [X_\alpha, Y_\alpha] \rangle &= \langle [x, X_\alpha], Y_\alpha \rangle \\ &= \alpha(x) \langle Y_\alpha, Y_\alpha \rangle, \end{aligned} \tag{8.16}$$

which gives us that $[X_\alpha, Y_\alpha] = \langle Y_\alpha, Y_\alpha \rangle v_\alpha$, when combined with the observation: $[X_\alpha, Y_\alpha] \in T$. An easy calculation now gives

$$\begin{aligned} ad\, Y_\alpha v_\alpha &= \alpha(v_\alpha) X_\alpha, \\ (ad\, Y_\alpha)^2 v_\alpha &= -\alpha(v_\alpha) \langle Y_\alpha, Y_\alpha \rangle v_\alpha. \end{aligned} \tag{8.17}$$

For $x \in L_\alpha$ we have $[Y_\alpha, x] = 0$ and so, we see $Ad(\exp t Y_\alpha) : T \to T$ by the formula

$$Ad(\exp t Y_\alpha) x = x - \frac{2\alpha(x)}{\alpha(v_\alpha)} v_\alpha, \tag{8.18}$$

which is the same formula as that of r_α, the reflection in the hyperplane L_α. Since $Ad(\exp t Y_\alpha) : T \to T$ it is an element of the Weyl group and so $r_\alpha \in W$ which completes the proof of the theorem.

COROLLARY 8.13 *The Weyl group acts transitively on the Weyl chambers.*

Proof We have already seen in Theorem 8.11 that the Weyl group permutes the Weyl chambers. Let C_1 and C_2 be any two Weyl chambers and pick elements $x_1 \in C_1$ and $x_2 \in C_2$. Consider the line joining x_1 to x_2. If this line crosses L_α then it is clear that

$$|x_1 - x_2| > |x_1 - r_\alpha x_2|, \tag{8.19}$$

where $| \ |$ is the norm associated to the Killing form innerproduct. The function $|x_1 - wx_2|$ on W is real valued and since W is compact attains its minimum. From (8.21) it is clear that x_1 and wx_2 lie in the same Weyl chamber for this minimum point w. Thus, we have $wC_2 \cap C_1 \neq \phi$ and since W permutes the Weyl chambers, this gives $WC_2 = C_1$.

We now proceed to the task of putting an ordering on the roots. One common way to do this is called a lexicographical ordering. This is a total ordering obtained by, essentially, ordering the roots arbitrarily, subject to a few obvious constraints. Our approach is to be satisfied with only a partial ordering of the roots. Even this is not completely natural. However, it is more natural than the lexicographical ordering and is sufficient for all the results which we shall need.

Pick a Weyl chamber C and call this the positive chamber.

DEFINITION 8.14   An element $\theta \epsilon T^*$ is positive if $\theta(x) > 0$ for all $x \epsilon C$.

PROPOSITION 8.15   *This defines a partial ordering on T^*.*

Proof   Immediate.

The Killing form is an innerproduct on T, and therefore, it defines an isomorphism
$$\sigma : T \to T^*, \tag{8.20}$$
which satisfies
$$< x, y > = \sigma(x)y. \tag{8.21}$$
By means of this isomorphism, we can define a Killing form on T^*, an action of the Weyl group on T^* and Weyl chambers in T^*. Let the image of the positive Weyl chamber C under σ be D^0. The following result is immediate from the definitions.

LEMMA 8.16   *If $\varphi \epsilon D^0$, then $< \varphi, \theta > > 0$ for all positive θ .*

DEFINITION 8.17   The dominant chamber D is the closure of D^0.

DEFINITION 8.18   A weight λ of a representation is a highest weight, if $\theta - \lambda > 0$ or $\theta = \lambda$ for any weight θ of the same representation implies that $\theta = \lambda$.

THEOREM 8.19   *If λ is the highest weight of a representation, then $\lambda \epsilon D$.*

Proof   Since the Weyl group action is induced from conjugation W permutes the weights of a representation. Let λ be the highest weight of a representation and suppose that $\lambda \notin D$. We shall see that this leads to a contradiction. Since W acts transitively on the Weyl chambers for some $w \in W$ we have $w\lambda \in D$. Clearly $w\lambda \neq \lambda$. Let $\theta \in D^0$. Then by a similar arguement to that which gave (8.21), we see w can be chosen so

$$\langle w\lambda, \theta \rangle > \langle \lambda, \theta \rangle. \tag{8.22}$$

Thus, for $x \in C$ we have

$$(W\lambda - \lambda)(x) = \langle w\lambda, \sigma(x) \rangle - \langle \lambda, \sigma(x) \rangle > 0, \tag{8.23}$$

since $\sigma(x) \in D^0$. Thus $w\lambda - \lambda > 0$ which contradicts the assumption that λ is a highest weight of a representation. This proves that $\lambda \in D$.

Remark 8.20   A weight which is in D is usually called dominant.

The following theorem describes the relationship between irreducible representation and dominant weights.

THEOREM 8.21   a)   *If π is an irreducible representation, then π has a unique highest weight, and this weight has multiplicity one.*

   b)   *If π_1 and π_2 are both irreducible representations, then π_1 is equivalent to π_2 if and only if they have the same highest weight.*

   c)   *If $\lambda \in P \cap D$ then, there is an irreducible representation with highest weight λ.*

The proof of Theorem 8.19 is postponed until the next chapter. In Chapter 9, a proof is given after Theorem 9.8, which establishes a formula due to Hermann Weyl for the character of a representation. In part c) of the theorem, we assert the existence of a representation. It is the content of the Borel-Weil theorem to explicitly construct this representation.

We shall now describe the roots and weights of the classical groups. Before doing this, we need some notation and an easy lemma.

LEMMA 8.22   *Let $L : V \to V$ be a linear transformation such that there are two vectors $u, v \in V$ with $Lu = av$ and $Lv = bu$. Then $\sqrt{ab}$ and $-\sqrt{ab}$ are eigenvalues of L.*

Proof A simple calculation shows

$$L\left(\sqrt{b}\,u + \sqrt{a}\,v\right) = \sqrt{ab}\left(\sqrt{b}\,u + \sqrt{a}\,v\right)$$
$$L\left(\sqrt{b}\,u - \sqrt{a}\,v\right) = -\sqrt{ab}\left(\sqrt{b}\,u - \sqrt{a}\,v\right). \qquad (8.24)$$

Notation 8.23 *The notation which we need is that, if* $\mathbf{R}^n = (\theta_1, \ldots, \theta_n)^t$, *then the element of the dual space* $\mathbf{R}^{n*}$ *given by* $(\theta_1, \ldots, \theta_n)^t \to \theta_i$ *is denoted by* θ_i^*.

Example 8.24 The roots and weights and Weyl group of $SU(n)$. As we saw in Chapter 6, the maximal torus is

$$T = \left\{ \begin{pmatrix} e^{i\theta_1} & & \\ & \ddots & \\ & & e^{i\theta_n} \end{pmatrix} : \theta_1 + \ldots + \theta_n = 0 \right\}. \qquad (8.25)$$

Thus, the Lie algebra to T is

$$\mathcal{T} = \left\{ \begin{pmatrix} i\theta_1 & & \\ & \ddots & \\ & & i\theta_n \end{pmatrix} : \theta_1 + \ldots + \theta_n = 0 \right\}. \qquad (8.26)$$

We identify this with $\mathbf{R}^n$ by

$$\begin{pmatrix} i\theta_1 & & \\ & \ddots & \\ & & i\theta_n \end{pmatrix} \to \begin{pmatrix} \theta_1 \\ \vdots \\ \theta_n \end{pmatrix}. \qquad (8.27)$$

From (8.2), the roots are now i times the eigenvalues of adjoint representation. Let

$$L = ad \begin{pmatrix} i\theta_1 & & \\ & \ddots & \\ & & i\theta_n \end{pmatrix}. \qquad (8.28)$$

Then, by (5.21), using $u = E_{ij}$ $v = F_{ij}$ in Lemma 8.22, we see the eigenvalues of L are

$$\mathbf{i}\,(\theta_i - \theta_j) \text{ and } -\mathbf{i}\,(\theta_i - \theta_j). \qquad (8.29)$$

Thus, identifying $\mathcal{T}$ with $\mathbf{R}^n$ by (8.29), and using Notation 8.23, we see the roots of $SU(n)$ are

$$\pm \left(\theta_i^* - \theta_j^*\right) \text{ for } i < j. \qquad (8.30)$$

The positive roots may be taken as

$$\theta_i^* - \theta_j^*, \tag{8.31}$$

and the innerproduct on T^* induced by the Killing form is, from (5.27), given by

$$< \sum a_i \theta_i^*, \sum b_i \theta_i^* > = \frac{1}{2n} \sum a_i b_i. \tag{8.32}$$

One-half the sum of the positive roots is

$$\rho = \frac{n-1}{2} \theta_1^* + \frac{n-3}{2} \theta_2^* + \ldots + \frac{3-n}{2} \theta_{n-1}^* + \frac{1-n}{2} \theta_n^*. \tag{8.33}$$

The integer lattice $I = \mathbf{z}^n$ in $\mathbf{R}^n$ with the sum of the coordinates zero. The lattice of roots $Q = \mathbf{z}^{n*}$ in $\mathbf{R}^{n*}$ with the sum of the coordinates equal to zero. Now, we identify the lattice of weights $P \subset \mathbf{R}^{n*}$. Clearly, elements of P have the sum of their coordinates equal to zero. It is equally clear that $Q \subset P$. Let

$$\gamma^* = \theta_1^* - \frac{1}{n}(\theta_1^* + \ldots + \theta_n^*). \tag{8.34}$$

Then, a simple calculation gives

$$\gamma^*(a_1, \ldots, a_n) = a_1 - \frac{1}{n}(a_1 + \ldots + a_n). \tag{8.35}$$

Now, if $(a_1, \ldots, a_n) \in I$ we have $a_1 \in \mathbf{z}$ and $a_1 + \ldots + a_n = 0$. Thus, we see $\gamma^* \in P$ and, in fact, P is generated by Q and γ^* (see Exercise 8.5). The dominant weights are those weights $a_1 \theta_1^* + \ldots + a_n \theta_n^*$ such that:

$$a_1 \geq a_2 \geq \ldots \geq a_n. \tag{8.36}$$

Let A be a permutation matrix, that is, one obtained by permuting the rows of the identity $n \times n$ matrix. Then,

$$A \begin{pmatrix} e^{i\theta_1} & & \\ & \ddots & \\ & & e^{i\theta_n} \end{pmatrix} A^{-1} = \begin{pmatrix} e^{i\theta_{\sigma(1)}} & & \\ & \ddots & \\ & & e^{i\theta_{\sigma(n)}} \end{pmatrix} \tag{8.37}$$

for a permutation σ and therefore, $A \in N(T)$. These A form a set of coset representatives for T in $N(T)$ and therefore, the Weyl group $W = S_n$ the symmetric group on n letters. The action of W on T and T^* is by permuting the coordinates of $\mathbf{R}^n$ and $\mathbf{R}^{n*}$.

Example 8.25   The roots, weights, and Weyl group of $SO(2n+1)$.

From Chapter 6, we know that the Lie algebra to the maximal torus is

$$\mathcal{T} = \left\{ \begin{pmatrix} \begin{matrix} 0 & \theta_1 \\ -\theta_1 & 0 \end{matrix} & & \\ & \ddots & \\ & & 0 \end{pmatrix} \right\}. \tag{8.38}$$

Thus, we identify $\mathcal{T}$ with $\mathbf{R}^n$ by

$$\begin{pmatrix} \begin{matrix} 0 & \theta_1 \\ -\theta_1 & 0 \end{matrix} & & \\ & \ddots & \\ & & 0 \end{pmatrix} \rightarrow \begin{pmatrix} \theta_1 \\ \vdots \\ \theta_n \end{pmatrix}. \tag{8.39}$$

From (5.35), if $L = ad\,(\theta_1, \ldots, \theta_n)^t$ and, using the notation of Example 5.11 in Chapter 5, we see that:

$$\begin{aligned}
L\,E_{2i,2j} &= \theta_i\,E_{2i-1,2j} + \theta_j\,E_{2i,2j-1}, \\
L\,E_{2i,2j-1} &= \theta_i\,E_{2i-1,2j-1} - \theta_j\,E_{2i,2j}, \\
L\,E_{2i-1,2j} &= -\theta_i\,E_{2i,2j} + \theta_j\,E_{2i-1,2j-1}, \\
L\,E_{2i-1,2j-1} &= -\theta_i\,E_{2i,2j-1} - \theta_j\,E_{2i-1,2j}. \tag{8.40}
\end{aligned}$$

Thus, for $1 \leq i < j \leq n$ the space V_{ij} spanned by $E_{2i,2j}, E_{2i,2j-1}, E_{2i-1,2j}$ and $E_{2i-1,2j-1}$ is invariant under L. The eigenvalues of L on V_{ij} are

$$\pm\,\mathbf{i}\,(\theta_i \pm \theta_j). \tag{8.41}$$

Notice that $E_{2i-1,2i}$ is in $\mathcal{T}$. The only other vectors in the Lie algebra $SO(2n+1)$ are $E_{2i-1,2n+1}$ and $E_{2i,2n+1}$. On these, we have

$$\begin{aligned}
L\,E_{2i-1,2n+1} &= -\theta_i E_{2i,2n+1}, \\
L\,E_{2i,2n+1} &= \theta_i E_{2i-1,2n+1}. \tag{8.42}
\end{aligned}$$

The eigenvalues of L on the space spanned by $E_{2i-1,2n+1}$ and $E_{2i,2n+1}$ are, therefore,

$$\pm\,\mathbf{i}\,\theta_i. \tag{8.43}$$

Since the roots are i times the eigenvalues of L, we have the roots of $SO(2n+1)$ are

$$\pm\,\theta_i^* \text{ and } \pm\left(\theta_i^* \pm \theta_j^*\right)\,(i < j). \tag{8.44}$$

The positive roots may be taken as θ_i^* and $\theta_i^* \pm \theta_j^*(i < j)$. The innerproduct induced from the Killing for is, from (5.37),

$$< \sum a_i \theta_i^*, \sum b_i \theta_i^* > = \frac{1}{(4n-2)} \sum a_i b_i. \tag{8.45}$$

One-half the sum of the positive roots is

$$\rho = \frac{(2n-1)\theta_1^*}{2} + \frac{(2n-3)}{2}\theta_2^* + \ldots + \frac{3\theta_{n-1}^*}{2} + \frac{\theta_n^*}{2}. \tag{8.46}$$

The integer lattice is $I = \mathbf{z}^n$ in $\mathbf{R}^n$ and the lattice of weights is $P = \mathbf{z}^{n*}$ in $\mathbf{R}^{n*}$. The Weyl group acts by permuting the $\theta_1^*, \ldots, \theta_n^*$ and by changing the signs of $\theta_1^*, \ldots, \theta_n^*$. It is, therefore, the semi-direct product of S_n with $\mathbf{z}_2^n$.

Remark 8.26 The group $SO(2n+1)$ is not simply connected. It does have a simply connected double cover $Spin(2n+1)$. The Lie algebra of $Spin(2n+1)$ is the same as that of $SO(2n+1)$, and therefore, the roots are the same. The integer lattice, however, has index two in the one given above. Thus,

$$I_{spin} = \mathbf{z}^n \text{ with the sum of the coordinates even.} \tag{8.47}$$

The lattice of weights for $Spin(2n+1)$ is then

$$P_{spin} = \mathbf{z}^{n*} + \mathbf{z}\left(\frac{1}{2}\sum \theta_i^*\right). \tag{8.48}$$

Notice that ρ, one-half the sum of the positive roots, is a weight for $Spin(2n+1)$ but, not for $SO(2n+1)$.

Example 8.27 The roots, weights, and Weyl group of $SO(2n)$. The Lie algebra of the maximal torus is given by

$$\mathcal{T} = \left\{\left(\begin{array}{cc} \begin{matrix} 0 & \theta_1 \\ -\theta_1 & 0 \end{matrix} & \\ & \ddots & \\ & & \begin{matrix} 0 & \theta_n \\ -\theta_n & 0 \end{matrix} \end{array}\right)\right\}. \tag{8.49}$$

We identify this in a way analogous to (8.41), with $\mathbf{R}^n = \{(\theta_1, \ldots, \theta_n)^t\}$. The equations analogous to (8.49) still hold, but there is nothing corresponding to (8.44). Thus, we see the roots are

$$\pm\left(\theta_i^* \pm \theta_j^*\right)(i < j). \tag{8.50}$$

The positive roots can be taken as $\theta_i^* \pm \theta_j^*$, $i < j$. Then, one-half of the sum of the positive roots is

$$\rho = (n-1)\theta_1^* + (n-2)\theta_2^* + \ldots + \theta_{n-1}^*. \qquad (8.51)$$

The innerproduct induced from the Killing form is

$$< \sum a_i \theta_i^*, \sum b_i \theta_i^* > = \frac{1}{4n-4} \sum a_i b_i. \qquad (8.52)$$

As in the previous example, the integer lattice is $I = \mathbf{z}^n$ in $\mathbf{R}^n$ and the lattice of weights is $P = \mathbf{z}^{n*}$ in $\mathbf{R}^{n*}$. The Weyl group acts by permuting $\theta_1^*, \ldots, \theta_n^*$ and by changing the sign of an even number of $\theta_1^*, \ldots, \theta_n^*$. It is, therefore, the semi-direct product of S_n and $\mathbf{z}_2^{n-1}$.

Remark 8.28   As in the previous example, $SO(2n)$ is not simply connected, but it does have a simply connected double cover $Spin(2n)$. The lattice of weights P_{so} then has index two in P_{spin} and

$$P_{spin} = \mathbf{z}^{n*} + \mathbf{z}\left(\frac{1}{2}\sum \theta_i^*\right). \qquad (8.53)$$

Example 8.29   The roots, weights, and Weyl group of $Sp(n)$.

The Lie algebra of the maximal torus of $Sp(n)$ is

$$\mathcal{T} = \left\{ \begin{pmatrix} \mathbf{i}\,\theta_1 & & \\ & \ddots & \\ & & \mathbf{i}\,\theta_n \end{pmatrix} \right\}. \qquad (8.54)$$

We identify this with $\mathbf{R}^n$ in a way analogous to (8.29). The equations in (5.29) give rise to the following three pairs of equations for $L = ad\,(\theta_1, \ldots, \theta_n)$. In each case, we can use Lemma 8.20 to write down the eigenvalues of L as follows:

$$\begin{aligned} L\,E_{ij} &= (\theta_i - \theta_j)\,F_{ij}, \\ \text{and} \quad L\,F_{ij} &= -(\theta_i - \theta_j)\,E_{ij}. \end{aligned} \qquad (8.55)$$

The eigenvalues of L are:

$$\pm\,\mathbf{i}\,(\theta_i - \theta_j); \qquad (8.56)$$

$$\begin{aligned} L\,G_{ij} &= (\theta_i + \theta_j)\,H_{ij}, \\ \text{and} \quad L\,H_{ij} &= -(\theta_i + \theta_j)\,G_{ij}. \end{aligned} \qquad (8.57)$$

The eigenvalues of L are:

$$\pm\, \mathbf{i}\,(\theta_i + \theta_j);\qquad\qquad (8.58)$$

$$\begin{aligned} L\,K_i &= 2\,\theta_i\,L_i,\\ L\,L_i &= -2\,\theta_i\,K_i. \end{aligned}\qquad\qquad (8.59)$$

The eigenvalues of L are:

$$\pm\, \mathbf{i}\,2\,\theta_i.\qquad\qquad (8.60)$$

Thus, the roots of $Sp(n)$ are

$$\pm\, 2\,\theta_i^*,\ \pm\left(\theta_i^* \pm \theta_j^*\right)\ (\text{ for } i < j).\qquad (8.61)$$

The innerproduct induced by the Killing form is

$$<\sum a_i\,\theta_i^*,\ \sum b_i\,\theta_i^*>\ =\ \frac{1}{4\,n+4}\ \sum a_i\,b_i.\qquad (8.62)$$

The positive roots may be taken as $2\,\theta_i^*$, $\theta_i^* \pm \theta_j^*$, for $i < j$. Then, one-half of the sum of the positive roots is

$$\rho = n\,\theta_1^* + (n-1)\,\theta_2^* +\ldots+ \theta_n^*.\qquad (8.63)$$

The integer lattice $I = \mathbf{z}^n$ in $\mathbf{R}^n$ and the lattice of weights is $P = \mathbf{z}^{n*}$ in $\mathbf{R}^{n*}$. The Weyl group acts by permuting the $\theta_1^*,\ldots,\theta_n^*$ and by changing the signs of $\theta_1^*,\ldots,\theta_n^*$. It is, thus, the semi-direct product of S_n and $\mathbf{z}_2^n$.

Exercises - Chapter 8

The group $SU(3)$ has three positive roots : $\alpha = \theta_1^* - \theta_2^*, \beta = \theta_2^* - \theta_3^*$ and $\rho = \theta_1^* - \theta_3^*$.

8.1 Show that one-half the sum of the positive roots of $SU(3)$ is ρ.

8.2 Show that the dominant weights of $SU(3)$ are $P \cap D = \{m\sigma + n\tau :$ $m, n \in \mathbf{z}, m \geq 0, n \geq 0\}$ where $\sigma = \frac{1}{3}\alpha + \frac{2}{3}\beta$ and $\tau = \frac{2}{3}\alpha + \frac{1}{3}\beta$. The group $Sp(2)$ has four positive roots: $\alpha = \theta_1^* + \theta_2^*, \beta = 2\,\theta_1^*, \gamma = 2\,\theta_2^*$ and $\delta = \theta_1^* - \theta_2^*$.

8.3 Show that one-half the sum of the positive roots of $Sp(2)$ is $\rho = \alpha + \frac{1}{2}\beta$.

8.4   Show that the dominant weights of $Sp(2)$ are $P \cap D$
$= \left\{ m\alpha + n \left(\frac{1}{2} \beta \right) : m, n \in \mathbf{z}, \ m \geq 0, \ n \geq 0 \right\}$.

8.5   In $SU(n)$, we saw, after equation (8.24), that if $\alpha = \sum \beta_i \theta_i^* + c\gamma^*$
with $\sum \beta_i = 0$, then $\alpha \in P$. Prove the converse : if $\alpha \in P$, then α has
this form. (Hint: Let $\alpha = \sum \alpha_i \theta_i^*$ and set $\alpha_i = \beta_i + \gamma_i$, $2 \leq i \leq n$,
with $\beta_i \in \mathbf{z}$, $0 \leq \gamma_i < 1$. Let $\beta_1 = -\beta_2 - \beta_2 - \ldots - \beta_n$, $\gamma_1 = \alpha_1 - \beta_1$ and
set $\gamma_1 = C + \delta$, $c \in \mathbf{z}$ and $0 \leq \delta < 1$. Then, show that $\gamma_i = \delta$, $2 \leq i \leq n$.
Now, use the fact that $\sum \gamma_i = 0$ to obtain the result.)

Let π_λ be the representation with the highest weight λ and π_μ with
the highest weight μ. Then, we can form the tensor product repre-
sentation $\pi_\lambda \otimes \pi_\mu$. For any representation $\pi : G \to Aut\ V$, if ν is a
weight, we define the weight space $V_{(\nu)} = \{v \in V : \pi(x)v = \nu(x)v$ for
all $x \in T\}$.

8.6   Let $\pi = \pi_\lambda \otimes \pi_\mu$ and show that ν is a weight of π if and only if
$\nu = \alpha + \beta$ where α is a weight of π_λ and β is a weight of π_μ.

8.7   Continue with $\pi = \pi_\lambda \otimes \pi_\mu$ as in Exercise 8.6. Show $\dim V_{(\nu)} =$
$\sum \dim V_{\lambda(\alpha)} \dim V_{\mu(\beta)}$ where the sum is over all pairs of weights α of
π_λ, β of π_μ such that $\alpha + \beta = \nu$ and $V_{\lambda(\alpha)}$ is the α weight space in V_λ.
8.8   From Exercise 8.7, show that $\pi_{\lambda + \mu}$ occurs with multiplicity one
as a direct summand of $\pi_\lambda \otimes \pi_\mu$.

8.9   Let $\theta^* = \frac{1}{2} (\theta_1^* - \theta_2^*)$ for the group $SU(2)$. Then, any weight of
$SU(2)$ is $n \theta^*$ and the dominant weights are those with $n \geq 0$. Show
the weights of $\pi_{n \theta^*}$ are $-n \theta^*$, $(-n + 2)\theta^*$, $\ldots$, $n \theta^*$.

8.10   Show how the results of Exercises 7.9 and 8.9 illustrate the
results of Exercises 8.6, 8.7, and 8.8.

Chapter 9

Weyl's Formulae

The aim of this chapter is to present four formulae established by Hermann Weyl. Two of these are particularly important: the integration formula and the character formula. The other two formulae, while not as important, are very useful. We begin this chapter with some notation and give the denominator formula. Then, we prove the integration formula, followed by the character formula. Finally, we give the dimension formula.

DEFINITION 9.1 Let $j(t) = \Sigma(-1)^\sigma \exp(i\sigma(\rho)t)$, with the sum over the Weyl group and ρ half the sum of the positive roots, be a function on T.

This function j is called the denominator function. The reason for this will be obvious when we come to the character formula. Notice that $\rho \epsilon P$. Therefore, j satisfies

$$j(t + 2\pi I) = j(t), \tag{9.1}$$

where I is the integer lattice. Thus, we can regard j as a function on T, the maximal torus of G.

Our first result is known as Weyl's denominator formula.

THEOREM 9.2 The function j is given by $j(t) = \Pi \left(\exp\left(\frac{i}{2}\alpha(t)\right) - \exp\left(\frac{-i}{2}\alpha(t)\right) \right)$ where the product is over all the positive roots α.

Proof Choose a basis $t_1^*, \ldots, t_\ell^*$ for T^* such that $t_k^* \in \frac{1}{2}P$.

Then, define $X_k(t)$ by

$$X_k(t) = \exp\left(it_k(t)\right). \tag{9.2}$$

Now $j(t)$ is skew invariant under the action of the Weyl group, that is, for $\sigma \in W$:

$$j(\sigma\, t) = (-1)^\sigma j(t). \tag{9.3}$$

Thus j is a skew invariant polynomial in X_k.

Now, consider the function

$$\Pi(t) = \Pi_{\alpha > 0}\left(\exp\left(\frac{i}{2}\,\alpha(t)\right) - \exp\left(\frac{-i}{2}\,\alpha(t)\right)\right). \tag{9.4}$$

This function is also a skew invariant polynomial in X_k. Clearly, by skew invariance Π divides j.

To complete the proof, we calculate the leading terms. For Π, we have

$$\Pi(t) = \exp(i\rho(t))\,\Pi_{\alpha > 0}(1 - \exp(i\,\alpha(t))). \tag{9.5}$$

Therefore, the leading term is $\exp(2\,\pi\,i\,\rho(t))$. This is also the leading term of j. Thus, Π divides j and has the same leading term, so that $\Pi = j$, which completes the proof.

The next main result is the Weyl integration formula. This is obtained as a corollary of the next theorem. However, before we can prove this theorem, we need a lemma and a definition.

DEFINITION 9.3   Let T_{reg} be the regular points of T.

LEMMA 9.4   *Let* $f : (G/T) \times T \to G$ *be* $f(sT, t) = sts^{-1}$. *Then,* $f \mid (G/T) \times T_{reg}$ *is a* $\mid W \mid$ *sheeted covering of an open dense set in* G.

Proof   We start by considering the map $\tilde{f} : G \times T \to G$:

$$\tilde{f}(s, t) = sts^{-1}. \tag{9.6}$$

The claim is that if t is regular, then $\tilde{f}$ has maximal rank. To see this, notice that

$$sts^{-1} = s_0\left(s_0^{-1}s\right)t\left(s_0^{-1}s\right)^{-1}s_0^{-1}. \tag{9.7}$$

Now, since $x \to s_0 x s_0^{-1}$ is an automorphism of G, we see that we need only check our claim for one s. Take $s = 1$, the identity element of G, and then, we see that f has maximal rank.

Next, we proceed by taking $\mathcal{M}$ to be an $ad(t)$ invariant subspace of $\mathcal{G}$, which is complementary to $\mathcal{T}$. For example,

$$\mathcal{G} = \mathcal{M} \oplus \mathcal{T}. \tag{9.8}$$

Therefore, we have, essentially, $\mathcal{M} = T_e(G/T)$. Consider the map $f_* : \mathcal{G} \times T \to G$, which is the lifting of $\tilde{f}$ to the Lie algebra given by:

$$f_* : (u, t) \to (\exp u)t(\exp u)^{-1}. \tag{9.9}$$

When we differentiate f_* we obtain the map

$$df_{*(l,t)} : \mathcal{G} \times \mathcal{T} \to \mathcal{G} \tag{9.10}$$

which is

$$df_{*(l,t)}(u, w) = \left(Ad(t)^{-1}u\right) - u + w. \tag{9.11}$$

In (9.11), we have used left translation to identify the tangent spaces with the Lie algebras. If $df_{*(l,t)}(u, w) = 0$, then we have $w \in \mathcal{M} \cap \mathcal{T}$, and therefore, $w = 0$. Thus, $u = Ad(t)^{-1}u$, and since T is regular $u = 0$. Thus, ker $df_{*(l,t)} = 0$, and therefore, rank $\tilde{f} = \dim(G/T) + \dim T = \dim G$, which is the greatest possible value for the rank.

To complete the proof, notice that $\tilde{f}$ is W invariant, and therefore, $\tilde{f}$ factors through $((G/T) \times T)/W$. Now $T - T_{reg}$ has measure zero, and therefore, by Sard's theorem, $f\left((G/T) \times (T - T_{reg})\right)$ has measure zero.

The main theorem follows:

THEOREM 9.5   *Let dm, dg and dt be the normalized measures induced by the Killing form on G/T, G and T, respectively. Then, $dg = \frac{|j(t)|^2}{|W|} \, dm \, dt$.*

Proof   This follows from the previous theorem when we calculate the Jacobian of f. By left translation, we see that this Jacobian does not depend upon m. Now, from (9.11), we see that

$$df(u, w) = (Adt)^{-1}u - u + w. \tag{9.12}$$

Thus

$$Jf = \det\left((Adt)^{-1} \mid \mathcal{M} - I\right), \tag{9.13}$$

but, $(Adt)^{-1}$ consists of 2×2 blocks of the form

$$\begin{pmatrix} \cos \alpha(t) & \sin \alpha(t) \\ -\sin \alpha(t) & \cos \alpha(t) \end{pmatrix}. \tag{9.14}$$

Since there is one block for each positive root α, we have

$$\begin{aligned}
Jf &= \Pi_{\alpha > 0} \left((\cos \alpha(t) - 1)^2 + \sin^2 \alpha(t) \right) \\
&= \Pi_{\alpha > 0} \left(2 - 2 \cos \alpha(t) \right) \\
&= \Pi_{\alpha > 0} \, 4 \sin^2 \frac{1}{2} \alpha(t) \\
&= \Pi_{\alpha > 0} \mid \exp \left(\frac{i}{2} \alpha(t) \right) - \exp \left(\frac{-i}{2} \alpha(t) \right) \mid^2 . \tag{9.15}
\end{aligned}$$

By the denominator formula, the proof is complete.

As a corollary to this, we state the Weyl integration formula. This is an immediate consequence of the previous theorem and requires no further proof.

COROLLARY 9.6   *If f is a central function on G, then*

$$\int_G f(g) dg = \frac{1}{|W|} \int_T f(t) \mid j(t) \mid^2 dt.$$

In the two preceeding results, the notation $|W|$ means the order of W. This is only well defined in this context if the Weyl group is a finite group. Recall that $W = N(T)/T$ and since $N(T)$ is a closed subgroup of the compact group G it is clear that W is compact. Until this point, the compactness of W has been sufficient for our needs, but now we prove the finiteness of W.

LEMMA 9.7   *The Weyl group W is finite.*

Proof   As we have seen W is compact. Thus, to prove W is finite it is sufficient to show that W is discrete. To prove this, it is sufficient to show that $N(T)$ and T have the same Lie algebra, $\mathcal{T}$. Let Y be an element of the Lie algebra of $N(T)$. Then, for any $x \in \mathcal{T}$, $[x, Y] \in \mathcal{T}$. Thus

$$\langle ad\, x(Y), ad\, x(Y) \rangle = -\langle (ad\, x)^2 Y, Y \rangle = 0, \tag{9.16}$$

since $ad\, x(Y) \in \mathcal{T}$ and $ad\, x|\mathcal{T} = 0$. That is $[x, Y] = 0$ and so $Y \in \mathcal{T}$ which completes the proof.

We now proceed to the character formula. First, we define some notation. Then, we prove the result.

DEFINITION 9.8 Let $j_\lambda(t) = \Sigma(-1)^\sigma \exp(i\sigma(\lambda+\rho)t))$, where the sum is over σ in the Weyl group, be a function on T. We define j_λ on T by the formula $j_\lambda(\exp t) = j_\lambda(t)$ and notice that the j_0 is the denominator function of Definition 9.1.

The Weyl character formula is the formula of the next theorem.

THEOREM 9.9 *Let χ_λ be the character of an irreducible representation of G with the highest weight λ. Then, $\chi_\lambda(t) = j_\lambda(t)/j(t)$ for $t \in T$.*

Proof Let g_λ be the central function on G, such that

$$g_\lambda \mid T = j_\lambda/j. \qquad (9.17)$$

Then $\{g_\lambda\}$ forms an orthonormal system in $L^2(G)$. First, observe

$$\int_G g_\lambda \bar{g}_\mu dg = \frac{1}{\mid W \mid} \int g_\lambda \bar{g}_m \mid j \mid^2 dt$$

$$= \frac{1}{\mid W \mid} \int j_\lambda \bar{j}_\mu dt. \qquad (9.18)$$

Now, if $\lambda \neq \mu$ then $\sigma(\lambda+\rho) - \rho$ are all distinct from $\sigma(\mu+\rho) - \rho$, and therefore, the integral in (9.18) is zero. Notice that we have removed a factor $\exp(i\rho(t))$ from both j. On the other hand, if $\lambda = \mu$ then $\exp(-i\rho(t))j_\lambda(t)$ is a combination of $\mid W \mid$ characters of T with coefficients ± 1. Thus, if $\lambda = \mu$ the integral is one. Now, we can write

$$\chi_\lambda \mid T = \sum n(\mu)g_\mu \qquad (9.19)$$

with the sum over dominant weights and $n(\mu) \geq 0$ for all μ. This follows from results about characters on T. However, we have from Theorem 7.12:

$$\int_G \mid \chi_\lambda \mid^2 dg = 1, \qquad (9.20)$$

since $\text{Hom}_G(V_\lambda, V_\lambda)$ is just the scalars, it has dimension one. Therefore, $n(\mu) = 0$ for all μ except one, when $n(\mu) = 1$. Thus, for some μ

$$\chi_\lambda = g_\mu. \qquad (9.21)$$

If we write $\chi_\lambda(t)$ and $g_\mu(t)$ as trigonometric polynomials, we find the leading terms are as follows:

the leading term of $\chi_\lambda(t) = \exp(i\lambda(t))$;

$$\text{the leading term of } g_\mu(t) = \exp(i\lambda(t)). \tag{9.22}$$

Thus, we see $\lambda = \mu$, which completes the proof of the theorem.

It is now clear why $j(t)$ is called the denominator function.

The proof of this result lets us give the proof of the highest weight theorem. This theorem was stated as Theorem 8.21.

Proof of Theorem 8.21   Let χ_λ and $\tilde{\chi}_\lambda$ both be characters of representations with highest weight λ. Then, these are both equal to g_λ, and therefore, we have

$$\chi_\lambda = \tilde{\chi}_\lambda. \tag{9.23}$$

It now follows from Remark 7.13, that up to equivalence there is only one irreducible representation with highest weight λ. Since the leading term of g_λ as a trigonometric polynomial is $\exp(i\lambda)$, the multiplicity of λ is one. Conversely, if λ is dominant, we can form g_λ. This can be written as a sum of characters of irreducible representations:

$$g_\lambda = \sum m(\mu)\chi_\mu. \tag{9.24}$$

Integrating and using (9.19) give

$$\sum m(\mu)^2 = 1. \tag{9.25}$$

Therefore, $m(\mu) = 0$ for all μ, except one for which $m(\mu) = 1$. Thus, $g_\lambda = \chi_\mu$, and therefore, $\lambda = \mu$ and the corresponding representation has highest weight λ.

As an application of this, we prove the Weyl dimension formula. Let V_λ be the representation space of the irreducible representation of highest weight λ. Then, $\dim V_\lambda = \chi_\lambda(1)$ and the Weyl dimension formula is an expression for this number.

THEOREM 9.10   *The dimension of V_λ is*

$$\dim V_\lambda = \Pi_{\alpha>0} < \lambda + \rho, \alpha > / \Pi_{\alpha>0} < \rho, \alpha > .$$

Proof   Let $f : T^* \to T$ be the isomorphism given by the Killing form. Set $x = \exp t f(\rho)$ where t is a real variable. Then, we calculate $\chi_\lambda(x)$:

$$
\begin{aligned}
\chi_\lambda(x) &= j_\lambda(tf(\rho))/j(tf(\rho)) \\
&= \sum(-1)^\sigma \exp(i < \sigma(\lambda + \rho), \rho > t)/j(tf(\rho)) \\
&= \sum(-1)^\sigma \exp(i < \sigma(\rho), \lambda + \rho > t), /j(tf(\rho)) \\
&= j(tf(\lambda + \rho))/j(tf(\rho)). \tag{9.26}
\end{aligned}
$$

Now, writing μ for either $\lambda + \rho$ or ρ, we see that $j(tf(\mu))$ has the following power series expansion:

$$j(tf(\mu)) = \Pi_{\alpha > 0}\left(\exp\left(\frac{i}{2}t\alpha\left(f\left(\mu\right)\right)\right) - \exp\left(\frac{-i}{2}t\alpha\left(f\left(\mu\right)\right)\right)\right)$$
$$= \Pi_{\alpha > 0}\left(it < \alpha, \mu > +0\left(t^2\right)\right), \tag{9.27}$$

where we have used the denominator formula. We wish to evaluate $\chi_\lambda(1)$, which means calculating

$$\lim_{t \to 0} j(tf(\lambda + \rho))/j(tf(\rho)). \tag{9.28}$$

By L'Hopital's rule, we see

$$\lim_{t \to 0} \frac{j(tf(\lambda + \rho))}{j(tf(\rho))} = \frac{\Pi_{\alpha > 0}(2\pi i < \alpha, \lambda + \rho >)}{\Pi_{\alpha > 0}(2\pi i < \alpha, \rho >)}. \tag{9.29}$$

When we cancel the factors $2\pi i$, we obtain the result of the theorem.

To complete the theoretical portion of this chapter, we notice the following simple result. First, define

$$d(\lambda) = \Pi_{\alpha > 0} < \alpha, \lambda > /\Pi_{\alpha > 0} < \alpha, \rho >, \tag{9.30}$$

so $\dim V_\lambda = d(\lambda + \rho)$. The polynomial d is called the dimension polynomial. The result is that d is harmonic.

THEOREM 9.11 *Let $\triangle$ be the Laplacian on functions on T^* given by the Killing form. Then, $\triangle d = 0$.*

Proof As a polynomial $d(\lambda)$ is skew symmetric and divides any skew symmetric polynomial. Since $\triangle$ is invariant $\triangle d$ is skew invariant, and therefore, d divides $\triangle d$. Now,

$$\text{degree } (\triangle d) \leq \text{ degree } (d) - 2,$$

and therefore, $\triangle d = 0$.

Example 9.12 *The characters of $SU(2)$.*

From Example 8.24, we see that the maximal torus of $SU(2)$ is

$$T = \left\{\begin{pmatrix} e^{i\theta} & 0 \\ 0 & e^{-i\theta} \end{pmatrix}\right\}. \tag{9.31}$$

There is one positive root α :

$$\alpha = 2\,\theta^*, \tag{9.32}$$

and one-half the sum of the positive roots is, therefore,

$$\rho = \theta^*. \tag{9.33}$$

The Lie algebra of the maximal torus T is $\mathcal{T} \cong \mathbf{R}$ and the Weyl group $W = \{1, -1\}$ where -1 acts by multiplication on $\mathbf{R}$. The lattice of weights is

$$P = \mathbf{z}\,\theta^*. \tag{9.34}$$

Therefore, we can identify a dominant weight λ as a non-negative integer. The denominator function is

$$\begin{aligned}
j(\theta) &= \sum(-1)^\sigma \exp(i\sigma(\rho)(\theta)) \\
&= e^{i\theta} - e^{-i\theta},
\end{aligned} \tag{9.35}$$

and the function j_λ is

$$\begin{aligned}
j_\lambda(\theta) &= \sum(-1)^\sigma \exp(i\sigma(\lambda + \rho)(\theta)) \\
&= e^{i(n+1)\theta} - e^{-i(n+1)\theta},
\end{aligned} \tag{9.36}$$

where $n\theta^*(\theta) = n\,\theta$. Thus, by the Weyl character formula

$$\chi_\lambda(\theta) = \frac{e^{i(n+1)\theta} - e^{-i(n+1)\theta}}{e^{i\theta} - e^{-i\theta}}. \tag{9.37}$$

If we compare (9.36) with (7.21), we see that $\lambda = n$, which is the character of the representation of $SU(2)$ on the homogeneous polynomials of degree n in two variables. Similarly, the Weyl dimension formula gives

$$\begin{aligned}
\dim V_\lambda &= \frac{<\lambda + \rho, 2\,\theta^*>}{<\rho, 2\,\theta^*>} \\
&= \lambda + 1,
\end{aligned} \tag{9.38}$$

which, for $\lambda = n$, is the dimension of the space of homogeneous polynomials in two variables of degree n. This justifies the assertion made in Example 7.22 about the irreducibility of the representations of $SU(2)$.

Exercises - Chapter 9

In Exercises 9.1 and 9.2, use the notation given before Exercise 8.1.

9.1   Find the dimension of the representation of $SU(3)$ with highest weight $\lambda = m\sigma + n\tau$.

9.2   Find the character $\chi_\lambda \begin{pmatrix} e^{i\theta_1} & & \\ & e^{i\theta_2} & \\ & & e^{i\theta_3} \end{pmatrix}$, with $\theta_1 + \theta_2 + \theta_3 = 0$ and highest weight $\lambda = m\sigma + n\tau$.

In Exercises 9.3 and 9.4, use the notation given before Exercise 8.3.

9.3   Find the dimension of the representation of $Sp(2)$, with highest weight $\lambda = m\alpha + n\left(\frac{1}{2}\beta\right)$.

9.4   Find the character $\chi_\lambda \begin{pmatrix} e^{i\theta_1} & \\ & e^{i\theta_2} \end{pmatrix}$ of the representation with highest weight $\lambda = m\alpha + n\left(\frac{1}{2}\beta\right)$.

9.5   Use Weyl's character formula to show $\chi_\lambda\left(g^{-1}\right) = \overline{\chi_\lambda(g)}$.

9.6   Let f be a class function on G. Show that $f \mid T$ is a Weyl group invariant function on T, and $j(t)f(t)$ is a Weyl group skew invariant function on T.

9.7   From the theory of Fourier analysis, we know that any function g on the torus T can be written $g(t) = \Sigma\, b_v \exp\left(2\pi i < v, \tilde{t} >\right)$ where $\tilde{t} \in \mathcal{T}$, such that $\exp\left(2\pi i \tilde{t}\right) = t$ and the sum is over vectors v in the integer lattice. Use this and exercise (9.6) to show $j(t)f(t) = \Sigma\, a_\lambda j_\lambda(t)$ for any central function f.

9.8   Define convolution of two functions f and g by $f * g(x) = \int_G f(xy)g\left(y^{-1}\right) dy$. Show that if f and g are both class functions, so is $f * g$.

Chapter 10

Differential Operators on Compact Lie Groups

In this chapter, we study differential operators on compact Lie groups. This has already been done in a special case. First order operators are, essentially, vector fields. Thus, studying the Lie algebra of a Lie group corresponds to studying left invariant first order differential operators. We pursue this point of view of relating differential operators to algebraic objects related to the Lie group. The first such object is the universal enveloping algebra. We begin this chapter by constructing this.

Let $\mathcal{G}$ be the Lie algebra of G. Then, the tensor algebra is

$$T\mathcal{G} = R \oplus \mathcal{G} \oplus (\mathcal{G} \otimes \mathcal{G}) \oplus \cdots; \qquad (10.1)$$

see Appendix 1. Inside $T\mathcal{G}$ let $\mathcal{S}$ be the ideal generated by the following terms

$$X \otimes Y - Y \otimes X - [X, Y] \qquad (10.2)$$

for all X and $Y \, \epsilon \, \mathcal{G}$. The universal enveloping algebra is then the quotient of these.

DEFINITION 10.1 Let $U = T\mathcal{G}/\mathcal{S}$. Then, the U is the universal enveloping algebra.

As stated above, there is a natural isomorphism between the Lie algebra of G and the left invariant first order differential operators on G. This is related to the following result, which is one version of the Poincaré-Birkhoff-Witt theorem.

THEOREM 10.2 *The natural map $i : \mathcal{G} \to U$ is an injective embedding.*

Proof   Let $\{X_1, \cdots, X_n\}$ be a basis for $\mathcal{G}$ and introduce the following set of vector spaces:

$$B_n = \text{linear span}\,(X_{j_1} \otimes \cdots \otimes X_{j_n}),\ j_1 < j_2 < \cdots < j_n, \qquad (10.3)$$

and

$$B = R \oplus B_1 \oplus B_2 \oplus \cdots. \qquad (10.4)$$

There is, of course, a natural map

$$B \to T\mathcal{G} \to U. \qquad (10.5)$$

First, we claim that this map is onto. The proof of this proceeds by induction on the degree of an element. Since $B_1 = \mathcal{G} = (T\mathcal{G})_1$ the result is clear for elements of degree one. The key step in the inductive proof is the following:

$$X_{j_1} \otimes \cdots \otimes X_{j_r} \otimes X_{j_{r+1}} \otimes \cdots \otimes X_{j_n} + \mathcal{S}$$
$$= X_{j_1} \otimes \cdots \otimes X_{j_{r+1}} \otimes X_{j_r} \otimes \cdots \otimes X_{j_n} \qquad (10.6)$$
$$+ X_{j_1} \otimes \cdots \otimes \left[X_{j_r}, X_{j_{r+1}}\right] \otimes \cdots \otimes X_{j_n} + \mathcal{S}.$$

This equation means that $mod\,\mathcal{S}$ and $mod\,T_1, \cdots, T_{n-1}$ the tensor product is commutative for elements of degree n. Here $T_k = \mathcal{G} \otimes \mathcal{G} \otimes \ldots \otimes \mathcal{G}$ where there is a total of k factors $\mathcal{G}$. This gives the natural grading on the tensor algebra $T\mathcal{G}$. Thus, by induction on n the composition (10.5) is onto.

Our second claim is that the map (10.5) has an inverse, σ. We define σ by induction on n. Let $u \epsilon U$ be represented by $X_{j_1} \otimes \cdots \otimes X_{j_n}$ in T. If $j_1 < j_2 < \cdots < j_n$ define σ by

$$\sigma(u) = X_{j_1} \otimes \cdots \otimes X_{j_n}. \qquad (10.7)$$

If $j_1 < j_2 < \cdots < j_n$ is not true, use (10.6) repeatedly to re-order $j_1, \cdots, j_n$ until it is true. Thus, we have

$$u = X_{j_1} \otimes \cdots \otimes X_{j_n} + t + s, \qquad (10.8)$$

where $t \epsilon T_{n-1}$ and $s \epsilon \mathcal{S}$. Then, define σ by

$$\sigma u = X_{j_1} \otimes \cdots \otimes X_{j_n} + \sigma t. \qquad (10.9)$$

This defines σ by induction. Clearly, we have arranged things so that $\sigma(\mathcal{S}) = 0$. Thus B is isomorphic to U. Now $\mathcal{G}$ is injectively embedded in B, therefore, $\mathcal{G}$ is injectively embedded in U. This completes the proof of the Poincaré-Birkhoff-Witt theorem.

Since the defining expressions of $\mathcal{S}$ (10.2) are always zero for an invariant differential operator, we can now identify U with the set of all left invariant differential operators. The Poincaré-Birkhoff-Witt theorem gives that all first order left invariant operators are in U. Any left invariant operator is now made up of first order operators and products of these. Thus, any left invariant operator is in U.

DEFINITION 10.3 Let $Z(U)$ be the center of U, that is, $X \epsilon Z(U)$ if and only if $XY = YX$ for all $Y \epsilon U$.

Note that $Z(U)$ is the set of bi-invariant, that is, both left and right invariant, differential operators on G. One operator in $Z(U)$ is particularly important. This operator is the Laplacian $\triangle$. Since left and right translation are both isometries, and the Laplacian commutes with isometries, it is clear that $\triangle$ is bi-invariant. We give another description below of $\triangle$ which clearly shows that $\triangle \epsilon Z(U)$.

The advantage of identifying differential operators with elements of the universal enveloping algebra is that we can use the algebraic structure to study differential equations. First, we see how a differential operator acts on a representation. Let

$$\pi : G \to Aut\, V \tag{10.10}$$

be a representation. Differentiate this to get

$$\pi : \to End\, V, \tag{10.11}$$

and therefore, by the universal property of U, we have the extension

$$\pi : U \to End\, V. \tag{10.12}$$

Thus, differential operators act on the elements of the representation space. Now, suppose that π is irreducible and $D \epsilon Z(U)$. Then, $\pi(D)$ acts on V and $\pi(D)$ commutes with the action of G since $D \in Z(u)$. Thus, by the Schur orthogonality relations $\pi(D)$ acts by scalar multiplication on V, that is,

$$\pi(D)v = p_D(\lambda)v \tag{10.13}$$

for all v in the irreducible representation of highest weight λ. In other words, we have a map

$$p_D : P \cap D \to \mathfrak{c}, \tag{10.14}$$

where the subscript D is a bi-invariant differential operator and the other D denotes the dominant cone. It turns out that this map p_D is a polynomial map. The result is the following theorem.

THEOREM 10.4 *If $C^W[T]$ is the space of Weyl group invariant polynomials on the Lie algebra T and $C^G[\mathcal{G}]$ the G invariant polynomials on $\mathcal{G}$, then these two spaces are isomorphic and $p_D : Z(U) \to C^W[T] \cong C^G[\mathcal{G}]$ is an isomorphism.*

Proof   Omitted.

Consider the following composition of maps:

$$\hom(\mathcal{G}, \mathcal{G}) \to \mathcal{G} \otimes \mathcal{G}^* \to \mathcal{G} \otimes \mathcal{G} \to U. \qquad (10.15)$$

These maps are as follows. The first is the natural isomorphism. The middle map is induced from the isomorphism $\mathcal{G}^* \to \mathcal{G}$ given by the Killing form, and the last map is the natural inclusion.

DEFINITION 10.5   The Casimir element is the image of the identity homomorphism under this composition.

If $X_1, \cdots, X_n$ is an orthonormal basis for $\mathcal{G}$ then, let $C = X_1^2 + \cdots + X_n^2 \in U$. Clearly C is the Casimir element. The polynomial $p_c(\lambda)$ can be calculated.

THEOREM 10.6   *The polynomial $p_c(\lambda)$ is*

$$p_c(\lambda) = \| \lambda + \rho \|^2 - \| \rho \|^2,$$

where ρ is one-half the sum of the positive roots and $\| \ \|^2$ is the square of the Killing form norm.

Proof   We make some special choices. First, for a basis of $\mathcal{G}$. Under the adjoint representation, $\mathcal{G}$ decomposes as

$$\mathcal{G} = T \oplus \sum_\alpha \mathcal{G}_\alpha, \qquad (10.16)$$

where $\mathcal{G}_\alpha$ is the two-dimensional subspace of $\mathcal{G}$ corresponding to the positive root α. For α pick e_α and $e_{-\alpha}$ so that for $x \in T$

$$Ad(x)e_\alpha = \alpha(x)e_\alpha \qquad (10.17)$$

and

$$Ad(x)e_{-\alpha} = -\alpha(x)e_{-\alpha}. \qquad (10.18)$$

Notice that e_α and $e_{-\alpha}$ are complex conjugates in the complexification of $\mathcal{G}_\alpha$. Normalize e_α and $e_{-\alpha}$ so they have unit length. Define h_α by

$$[e_\alpha, e_{-\alpha}] = h_\alpha. \tag{10.19}$$

The second choice is in the space V_λ. Pick v so that

$$\pi_\lambda(x)v = \lambda(x)v \tag{10.20}$$

for all $x \in T$. Then, we have

$$\pi_\lambda(C)v = p_c(\lambda)v \tag{10.21}$$

and

$$C = \sum h_i^2 + \sum e_\alpha e_{-\alpha} + e_{-\alpha}e_\alpha. \tag{10.22}$$

Thus, we can calculate $p_c(\lambda)$:

$$\begin{aligned} p_c(\lambda) &= \sum \pi_\lambda (h_i)^2 + \sum \pi_\lambda (2e_{-\alpha}e_\alpha + h_\alpha) \\ &= \| \lambda \|^2 + \sum < \lambda, \alpha > \\ &= \| \lambda \|^2 + < \lambda, 2\rho > \\ &= \| \lambda + \rho \|^2 - \| \rho \|^2 . \end{aligned} \tag{10.23}$$

This completes the proof of the theorem and concludes this chapter.

Exercises - Chapter 10

10.1   For the group $SU(2)$ the dominant weights are $n\rho$, $n \in \mathbf{z}$ with $n \geq 0$ and ρ one-half the sum of the positive roots. Show that the eigenvalues of the Casimir are $n(n + 2)/8$.

10.2   Using the notation given in Exercise 8.2 for the group $SU(3)$, show that for $\lambda = m\sigma + n\tau$ the eigenvalue of the Casimir is $(m^2 + 3m + mn + n^2 + 3n)/9$.

10.3   Let $\pi : G \to V$ be an irreducible representation of G and let $<, >$ be a G-invariant innerproduct on V. If $u, v \in V$ the matrix coefficient of π with respect to u and v is $f_{uv}(g) = < \pi(g)u, v >$. Show that f_{uv} is an eigenfunction for the Laplacian on G.

10.4   The representations of $SO(3)$ are given in Example 8.23, and have highest weights $n\,\theta^*$. Show that the eigenvalues of the Casimir are $\frac{1}{2}n(n + 1)$.

10.5   The group $SO(3)$ acts by isometries on S^2. Hence, the eigen-
spaces of the Laplacian on S^2 are representation spaces of $SO(3)$. Use
the result of Exercise 10.4 to show that if λ is an eigenvalue of the
Laplacian on S^2 with multiplicity m_λ then, for some n, $\lambda = \frac{1}{2}n(n+1)$
and m_λ is a multiple of $2n + 1$.

Chapter 11

The Riemannian Geometry of a Compact Lie Group

The compact Lie group G is a manifold. Its tangent bundle can be trivialized by using left translation:

$$T(G) = G \times \mathcal{G}. \tag{11.1}$$

If we restrict ourselves to the case when G is semisimple, the Killing form induces an innerproduct on $\mathcal{G}$:

$$< X, Y > = -tr(ad\, X\, ad\, Y). \tag{11.2}$$

Thus, using the trivialization (11.1), G has a natural Riemannian metric.

An affine connection on G is a map ∇, which assigns to any two vector fields X and Y a third vector field $\nabla_X Y$. This satisfies the following properties:

$$
\begin{array}{lll}
\text{i)} & \nabla_X(Y + Z) = \nabla_X Y + \nabla_X Z, & \\
\text{ii)} & \nabla_{X+Y}(Z) = \nabla_X Z + \nabla_Y Z, & \\
\text{iii)} & \nabla_X(fY) = X(f)Y + f\, \nabla_X Y, & \\
\text{iv)} & \nabla_{fX} Y = f\, \nabla_X Y, & (11.3)
\end{array}
$$

where X, Y and Z are vector fields and f is a C^∞ real valued function on G. Such a connection is compatible with the Riemannian metric on G if

$$X < Y, Z > = < \nabla_X Y, Z > + < Y, \nabla_X Z >, \tag{11.4}$$

for any three vector fields X, Y and Z. Notice that if Y and Z are vector fields $< Y, Z >$ is a real valued function on G, and hence, $X < Y, Z >$ is another real valued function on G. For a connection, the torsion T and curvature R are defined by

$$
\begin{array}{lll}
T(X,Y) & = & \nabla_X Y - \nabla_Y X - [X,Y] \\
\text{and } R(X,Y)Z & = & -\nabla_X \nabla_Y Z + \nabla_Y \nabla_X Z + \nabla_{[X,Y]} Z.
\end{array}
\tag{11.5}
$$

This definition of the curvature tensor agrees with that in [9]. Many authors give R the opposite sign. This sign convention has the advantage that in a system of local coordinates $< R\left(\frac{\partial}{\partial x^i}, \frac{\partial}{\partial x^j}\right)\frac{\partial}{\partial x^k}, \frac{\partial}{\partial x^\ell} >$ coincides with the classical $R_{ijk\ell}$.

THEOREM 11.1   *There is a unique connection which is compatible with the metric and has $T(X, Y) = 0$ for all X and Y.*

Proof   We work in local coordinates. Let $x^1, \ldots, x^n$ be a system of local coordinates. Set $X_i = \frac{\partial}{\partial x^i}$.

Then, the metric is given by

$$g_{ij} = < X_i, X_j > . \tag{11.6}$$

Since ∇ is compatible with the metric, we have

$$X_i g_{jk} = < \nabla_{X_i} X_j, X_k > + < X_j, \nabla_{X_i} X_k > . \tag{11.7}$$

Now, ∇ has zero torsion. Therefore, since $[X_i, X_j] = 0$ for coordinate vector fields, we have $\nabla_{X_i} X_j = \nabla_{X_j} X_i$.

Permuting the indices in (11.7) gives three linear equations for $< \nabla_{X_i} X_j, X_k >$, $< \nabla_{X_j} X_k, X_i >$ and $< \nabla_{X_k} X_i, X_j >$. These can be solved to give

$$< \nabla_{X_i} X_j, X_k > = \frac{1}{2}(X_i g_{jk} + X_j g_{ik} - X_k g_{ij}). \tag{11.8}$$

Equation (11.8) proves the uniqueness of ∇. Conversely, if we set

$$\Gamma_{ij}^k = \frac{1}{2}\sum(X_i g_{j\ell} + X_j g_{i\ell} - X_\ell g_{ij})\, g^{\ell k}, \tag{11.9}$$

where $\left(g^{\ell k}\right)$ is the inverse matrix to (g_{ij}), then we can define ∇ by

$$\nabla_{X_i} X_j = \sum \Gamma_{ij}^k X_k, \tag{11.10}$$

with the sum over the repeated index k. The existence of ∇ now follows from equation (11.9).

The connection which has just been defined is called the Levi-Civita connection. The functions Γ_{ij}^k are called Christoffel symbols, while equations (11.8) and (11.9) are known as the first and second

Christoffel identities.

If X, Y and Z are orthonormal vector fields, then they no longer come from coordinates and need not have a zero Lie bracket. However, by similar arguments to those above, one can show:

THEOREM 11.2 *For orthonormal vector fields*

$$< \nabla_X Y, Z > = \frac{1}{2} (< Z, [X, Y] > - < Y, [X, Z] > - < X, [Y, Z] >).$$

Proof   This is left to the reader as an exercise.

By using the trivialization (11.1), left invariant vector fields can be identified with elements of the Lie algebra. Thus, if we restrict the connection to left invariant fields, we obtain a connection

$$\nabla : \mathcal{G} \times \mathcal{G} \to \mathcal{G}. \tag{11.11}$$

For this restriction to left invariant vector fields, we have the following:

THEOREM 11.3 $\nabla_X Y = \frac{1}{2} [X, Y].$

Proof   From the left invariance of the innerproduct,

$$Z < X, Y > = < [Z, X], Y > + < X, [Z, Y] > . \tag{11.12}$$

Now $< X, Y >$ is constant on G, so $< Y, [X, Z] > = - < X, [Y, Z] >$. The result now follows from Theorem 11.2.

COROLLARY 11.4 $R(X, Y)Z = \frac{1}{4} [[X, Y], Z].$

The Riemannian curvature has many symmetries. We list some in the following lemma.

LEMMA 11.5

$$
\begin{array}{ll}
\text{i)} & R(X, Y)Z + R(Y, X)Z = 0, \\
\text{ii)} & < R(X, Y)Z, W > + < R(X, Y)W, Z > = 0, \\
\text{iii)} & R(X, Y)Z + R(Y, Z)X + R(Z, X)Y = 0, \\
\text{iv)} & < R(X, Y)Z, W > = < R(Z, W)X, Y > .
\end{array}
$$

Proof   In this case, these symmetries follow immediately upon substituting the formula of Corollary 11.4, and using the invariance of the inner product.

It is true that the symmetries of Lemma 11.5 hold on a general Riemannian manifold. In particular, iii) which follows in this case from the Jacobi identity, known as "the first Bianchi identity." These symmetries show that the Riemannian curvature is a complicated object. Hence, one would like to contract it to something more simple. Again, the symmetries come to our aid—this time by indicating the best way to proceed, that is, to form the Ricci curvature.

The Riemannian curvature tensor can be contracted in various ways. The most important and common of these are the sectional curvature $S(X, Y)$, the Ricci curvature $Ric(x)$, and the scalar curvature κ.

Let X and Y be two orthogonal unit vectors in $\mathcal{G}$. Then X and Y determine a plane in $\mathcal{G}$ and by left translation, a bundle of planes over G. At each point p of G, we have a local diffeomorphism

$$\exp : U_p \to N_p \qquad (11.13)$$

from a neighborhood U_p of the origin in $T_p(G)$ to a neighborhood N_p of p in G. Thus, the two-dimensional subspace of $T_p(G)$ given by X and Y determines (locally) a two-dimensional submanifold of G. The sectional curvature is a measure of the curvature of this submanifold.

DEFINITION 11.6 $S(X, Y) = <R(X,Y)X, Y>$.

LEMMA 11.7 $S(X, Y) = \frac{1}{4} <[X, Y], [X, Y]>$.

Proof   This is a calculation from equation (11.4):

$$\begin{aligned} S(X, Y) &= <R(X, Y)X, Y> \\ &= \frac{1}{4} <[[X, Y], X], Y>, \qquad (11.14) \end{aligned}$$

and the result follows from the invariance of the inner product.

DEFINITION 11.8   The Ricci curvature $r(X, Y)$ is the trace of the map $Z \to R(X, Z)Y$.

LEMMA 11.9 $r(X, Y) = \frac{1}{4} \Sigma <[X, E_i], [Y, E_i]>$ *where the sum is over an orthonormal basis* $\{E_i\}$ *for* $\mathcal{G}$.

Proof We calculate

$$
\begin{aligned}
r(X,Y) &= tr(Z \to R(X,Z)Y) \\
&= \sum < R((X,E_i)\,Y, E_i > \\
&= \frac{1}{4} \sum < [[X,E_i]\,Y]\,, E_i > \\
&= \frac{1}{4} \sum < [X,E_i]\,, [Y,E_i]\,, > .
\end{aligned}
\tag{11.15}
$$

Alternatively, instead of considering the Ricci curvature as a map $r : TM \times TM \to \mathbf{R}$ we can define it as a map $Ric : TM \to TM$. This is done by the formula:

$$
r(X,Y) = \, < Ric(X), Y > .
\tag{11.16}
$$

LEMMA 11.10 $Ric(X) = -\frac{1}{4} \sum [[X,E_i]\,E_i].$

Proof We check this formula by a calculation:

$$
\begin{aligned}
\left\langle \frac{1}{4} \sum [[X,E_i]\,E_i]\,, Y \right\rangle &= -\frac{1}{4} \sum \left\langle [[X,E_i]\,E_i]\,, Y \right\rangle \\
&\quad -\frac{1}{4} \sum \left\langle [X,E_i]\,, [Y,E_i] \right\rangle .
\end{aligned}
\tag{11.17}
$$

This is $r(X,Y)$ by the result of Lemma 11.9.

Finally, there is the scalar curvature. This is a complete contraction of the Riemannian curvature tensor.

DEFINITION 11.11 The scalar curvature $\kappa = trace\,Ric$.

LEMMA 11.12 $\kappa = \frac{1}{4} \dim G.$

Proof

$$
\begin{aligned}
\kappa &= trace\,Ric \\
&= \sum \left\langle Ric\,(E_i)\,, E_i \right\rangle \\
&= -\frac{1}{4} \sum \left\langle [[E_i, E_j]\,E_j]\,, E_i \right\rangle \\
&= \frac{1}{4} \sum \left\langle [E_i, E_j]\,, [E_i, E_j] \right\rangle \\
&= \frac{1}{4} \dim G.
\end{aligned}
\tag{11.18}
$$

The last step in the calculation which proves Lemma 11.12 has the following interpretation.

Theorem 11.13 $r(X,Y) = \frac{1}{4} <X,Y>$.

Proof   We recall the definition of the innerproduct:

$$
\begin{aligned}
<X,Y> &= -tr(ad\,X\,ad\,Y) \\
&= -\sum \Big\langle [X\,[Y,E_i]]\,,E_i \Big\rangle \\
&= \sum \Big\langle [Y,E_i]\,,[X,E_i] \Big\rangle \\
&= 4r(X,Y). \tag{11.19}
\end{aligned}
$$

Corollary 11.14   *G is an Einstein manifold.*

Exercises - Chapter 11

11.1   Give the details in obtaining equation (11.8):

$$
\Big\langle \nabla_{X_i} X_j, X_k \Big\rangle = \frac{1}{2} \left(X_i g_{jk} + X_j g_{ik} - X_k g_{ij} \right).
$$

11.2   Give the proof of Theorem 11.2: for orthonormal vector fields

$$
\Big\langle \nabla_X Y, Z \Big\rangle = \frac{1}{2} \left(<Z,[X,Y]> - <Y,[X,Z]> - <X,[Y,Z]> \right).
$$

11.3   Show how the invariance of the innerproduct gives the result

$$
< [[X,Y],X],Y > = < [X,Y],[X,Y] > .
$$

11.4   Use Lemma 11.12 to give the scalar curvature for the groups $SU(2)$, $SU(3)$, $SO(3)$, $SO(4)$ and $Sp(2)$.

Chapter 12

The Trace of the Heat Kernel

On any compact Riemannian manifold, let $\{\lambda, \phi_\lambda\}$ be a complete set of eigenvalues, λ, and corresponding eigenfunctions, ϕ_λ, for the Laplace-Beltrami operator. We fix the signs so that the eigenvalues form a positive increasing sequence

$$0 = \lambda_1 \leq \lambda_2 \leq \ldots . \tag{12.1}$$

This corresponds to $\Delta = -\dfrac{\partial^2}{\partial x^2} - \dfrac{\partial^2}{\partial y^2} - \dfrac{\partial^2}{\partial z^2}$ on subsets of $\mathbf{R}^3$ with the usual metric, and therefore, it is the opposite sign to that customarily used by analysts. Then, the Heat Equation is

$$\Delta u + \frac{\partial u}{\partial t} = 0, \tag{12.2}$$

and the Heat Kernel is

$$H(x, y, t) = \sum \phi_\lambda(x)\phi_\lambda(y)e^{-\lambda t}. \tag{12.3}$$

If $v(x)$ is a distribution, giving the initial data, the function

$$u(x, t) = \int H(x, y, t)v(y)dy \tag{12.4}$$

satisfies the Heat Equation and $u(x, 0) = v(x)$. The Heat Kernel is of trace class. Its trace is

$$Z(t) = \int H(x, x, t)dx. \tag{12.5}$$

Thus, we see

$$Z(t) = \sum_\lambda n_\lambda e^{-\lambda t} \tag{12.6}$$

where n_λ is the multiplicity of the eigenvalue λ. This function, $Z(t)$, carries a great deal of information about the geometry of the manifold.

In this book we are interested in the study of compact Lie groups, rather than more general Riemannian manifolds. In Chapter 9, we saw that the Casimir element of the universal enveloping algebra played the role of the Laplace-Beltrami operator. Its eigenvalues are:

$$c(\lambda) = \| \lambda + \rho \|^2 - \| \rho \|^2 . \tag{12.7}$$

The corresponding eigenfuctions are the matrix coefficients of the representation π_λ. Hence, the multiplicity is given by

$$n_\lambda = (\dim \pi_\lambda)^2 = d(\lambda + \rho)^2. \tag{12.8}$$

Thus, the trace of the Heat Kernel on a compact Lie group is

$$Z(t) = \sum_{\lambda \in P \cap D} d(\lambda + \rho)^2 \, e^{-(\|\lambda+\rho\|^2 - \|\rho\|^2)t}. \tag{12.9}$$

LEMMA 12.1 $Z(t) = \dfrac{e^{\|\rho\|^2 t}}{|W|} \Sigma_{\lambda \in P} \, d(\lambda)^2 \, e^{-\|\lambda\|^2 t}.$

Proof   Let us define $\theta(t)$ by

$$\theta(t) = \sum_{\lambda \in P \cap D} d(\lambda + \rho)^2 \, e^{-\|\lambda+\rho\|^2 t}. \tag{12.10}$$

Now, consider the map $\lambda \to \lambda + \rho$. Under this map, $P \cap D$ goes onto $P \cap D^\circ$, the interior points of the dominant Weyl chamber. Since d is skew symmetric under the Weyl group, we see $d(\lambda) = 0$ if λ is in the walls of the Weyl chamber. Thus,

$$\theta(t) = \sum_{\lambda \in P \cap D} d(\lambda)^2 \, e^{-\|\lambda\|^2 t}. \tag{12.11}$$

Now $\| \lambda \|^2$ and $d(\lambda)^2$ are invariant under the action of the Weyl group. Thus,

$$\theta(t) = \frac{1}{|W|} \sum_{\lambda \in P} d(\lambda)^2 \, e^{-\|\lambda\|^2 t}, \tag{12.12}$$

where $| W |$ denotes the order of the Weyl group. Some of the lattice points in the walls of the Weyl chamber are counted more than once in this averaging procedure. However, since $d(\lambda) = 0$ for such points, the sum is not changed. Since

$$Z(t) = e^{\|\rho\|^2 t} \theta(t) \tag{12.13}$$

the proof of Lemma 12.1 is complete.

We shall use the results of Lemma 12.1 to calculate the asymptotic expansion of $Z(t)$ as $t \to 0$. The basic method is to use the Fourier transform and Poisson summation formula. Since this material is standard to the subject of Fourier analysis, we shall only outline the main steps. The details are left to the reader as exercises.

DEFINITION 12.2   The Fourier transform is defined by

$$\hat{f}(\xi) = \int_{-\infty}^{\infty} e^{2\pi <\xi,\lambda>} f(\lambda) d\lambda.$$

The Poisson summation formula is

$$\sum_{\lambda \in L} f(\lambda) = \frac{1}{vol\, L} \sum_{\xi \in L^*} \hat{f}(\xi). \tag{12.14}$$

In (12.14), L is a lattice, *vol* L is the volume of a fundamental parallelepiped of L and L^* denotes the lattice dual to L.

LEMMA 12.3   *If* $g(\lambda) = e^{-\|\lambda\|^2 t}$, *then the Fourier transform is* $\hat{g}(\xi) = \left(\frac{t}{\pi}\right)^{-\ell/2} e^{-\pi^2 \|\xi\|^2 / t}$.

Proof   This is a direct calculation:

$$\hat{g}(\xi) = \int_{-\infty}^{\infty} e^{2\pi i <\xi,\lambda>} e^{-\|\lambda\|^2 t} d\lambda, \tag{12.15}$$

which is left to the reader as an exercise.

If $p(\lambda)$ is a polynomial and $f(\lambda) = p(\lambda)g(\lambda)$, then by differentiating under the integral sign, we see

$$\hat{f}(\xi) = p\left(\frac{1}{2\pi i} \frac{\partial}{\partial \xi}\right) \hat{g}(\xi). \tag{12.16}$$

We apply this when $p(\lambda) = d(\lambda)^2$ and $g(\lambda) = e^{-\|\xi\|^2 t}$. Here, we find

$$p\left(\frac{1}{2\pi i} \frac{\partial}{\partial \xi}\right) \hat{g}(\xi) = \mathcal{H}d^2(\xi) \left(\frac{t}{\pi}\right)^{-\ell/2} e^{-\pi^2 \|\xi\|^2 / t}. \tag{12.17}$$

The first term $\mathcal{H}d^2(\xi)$ is a polynomial in ξ. The symbol $\mathcal{H}$ is used because of the relation with the Hermite polynomials, which is explained below.

DEFINITION 12.4   The kth Hermite polynomial with parameter u is
$h_k(x, u) = e^{-ux^2/2} \left(\frac{d}{dx}\right)^k e^{ux^2/2}$.

LEMMA 12.5 $h_k(x, u) = u^k \sum_{r=0}^{\infty} \left(\frac{d}{dx}\right)^{2r} x^k \frac{u^r}{2^r r!}$.

Proof   For $k = 0$ and $k = 1$ this is clearly true. Now, from Definition 12.4, we see that h_k satisfies the recurrence relation

$$h_{k+1}(x, u) = \frac{d}{dx} h_k(x, u) + uxh_k(x, u). \qquad (12.18)$$

It is easy to see that the expression in Lemma 12.5 satisfies the same recurrence relation.

We are interested in the case when $p(\lambda)$ is homogeneous of degree $2n$, where n is the number of positive roots of G. Using multi-indices, we have $p(\lambda) = \sum a_\alpha \lambda^\alpha$. Then, we see

$$\mathcal{H}d^2(\xi) = (2\pi i)^{-2n} \sum a_\alpha h_\alpha(\xi, u), \qquad (12.19)$$

where

$$h_\alpha(\xi, u) = h_{\alpha_1}(\xi_1, u) h_{\alpha_2}(\xi_2, u) \ldots h_{\alpha_\ell}(\xi_\ell, u), \qquad (12.20)$$

and $u = -2\pi^2/t$. Formally, we can write Lemma 12.5 as

$$h_k(x, u) = \exp\left(\frac{u\Delta}{2}\right)\left(u^k x^k\right), \qquad (12.21)$$

where $\Delta = \frac{d^2}{dx^2}$. This lets us write equation (12.19) as

$$\mathcal{H}d^2(\xi) = (2\pi i)^{-2n} \exp\left(\frac{u\Delta}{2}\right) d^2(\xi) u^{2n}, \qquad (12.22)$$

where, this time, Δ is the Laplacian in ξ.

We can now obtain the asymptotic expansion of $Z(t)$ as $t \to 0$. First, we observe that using equations (12.16) and (12.17), and the Poisson summation formula

$$\theta(t) = \frac{1}{|W| \, vol \, L} \sum_{\xi \in p^*} \mathcal{H}d^2(\xi) \left(\frac{t}{\pi}\right)^{-\ell/2} e^{-\pi^2 \|\xi\|^2/t}. \qquad (12.23)$$

Thus, as $t \to 0$, since $e^{-\pi^2 \|\xi\|^2 / t}$ has a zero asymptotic expansion for $\xi \neq 0$, we see

$$\theta(t) \sim \frac{1}{|W| \, vol \, L} \left(\frac{t}{\pi}\right)^{-\ell/2} \mathcal{H}d^2(0). \qquad (12.24)$$

This shows us that there is a constant C such that

$$\theta(t) \sim C \, t^{-\dim G/2}, \qquad (12.25)$$

where we have used $\dim G = 2n + \ell$.

LEMMA 12.6 $Z(t) \sim C \, t^{-\dim G/2} e^{\|\rho\|^2 t}$.

Proof   This is immediate since $Z(t) = e^{\|\rho\|^2 t} \theta(t)$.

For a compact Riemannian manifold, M, the trace of the Heat Kernel has an asymptotic expansion. This is well-known to be

$$Z(t) = (4\pi t)^{-n/2} vol \, M \left(1 + \kappa t/6 + 0\left(t^2\right)\right), \qquad (12.26)$$

where $n = \dim M$ and κ is the scalar curvature. Comparing this with the result of Lemma 12.6 gives

$$\frac{\kappa}{6} = \| \rho \|^2 \qquad (12.27)$$

for a compact Lie group G. Thus, by Lemma (11.12), we see

$$\| \rho \|^2 = \frac{\dim G}{24}. \qquad (12.28)$$

This is the "strange formula" of Freudenthal and deVries. If we had been more careful of the details in calculating the constant C in equation (12.25), we would have obtained an expression for the Riemannian volume of G. The result follows.

THEOREM 12.7 *The volume of G is given by $volG = (2\pi)^{\ell+n} volQ^v / \Pi_{\alpha>0}$*

$< \alpha, \rho >$, *where $vol \, Q^v$ is the volume of a fundamental parallelepiped of the lattice generated by the coroots of G with respect to the inner-product induced by the Killing form.*

Exercises - Chapter 12

12.1   Complete the proof of Lemma 12.3. That is, for $g(\lambda) = e^{-\|\lambda\|^2 t}$ calculate the Fourier transform $\hat{g}(\xi) = \int_{-\infty}^{\infty} e^{2\pi<\xi,\lambda>} g(\lambda)d\lambda$.

12.2   Check that $u^k \sum_{r=0}^{\infty} \left(\frac{d}{dx}\right)^{2r} \frac{x^k u^r}{2^r r!} = f_k(x, u)$ satisfies the recurrence relation (12.18) : $f_{k+1}(x, u) = \frac{d}{dx} f_k(x, u) + ux f_k(x, u)$.

12.3   Calculate $\| \rho \|^2$ and $\dim G/24$, explicitly, for the groups $SU(2), SU(3), Spin(4)$ and $Sp(2)$. Check to see if this verifies the "strange formula" equation (12.28) in the case of these four groups.

12.4   (Harder) Give the details of the proof of Theorem 12.7 by identifying the constant C of equation (12.25).

Appendix 1: The Tensor Product

Let V and W be two vector spaces. Then, the Cartesian product is $V \times W = \{(v, w) : v \in V, w \in W\}$. The Cartesian product is a vector space with operations

$$\begin{aligned} (v_1, w_1) + (v_2, w_2) &= (v_1 + v_2, w_1 + w_2), \\ \text{and} \quad r(v, w) &= (rv, rw). \end{aligned} \tag{A1.1}$$

A map $T : V \times W \to U$ is bilinear if it is linear in each factor.

DEFINITION A1.1 The tensor product of V and W is a vector space $V \otimes W$ and a bilinear map $i : V \times W \to V \otimes W$ such that for any bilinear map $T : V \times W \to U$ there is a unique linear map $g : V \otimes W \to U$ with $T = g \circ i$.

It is not clear from this definition whether tensor products exist. However, it is easy to see that if tensor products exist, they are unique up to isomorphism.

PROPOSITION A1.2 *Let $j : V \times W \to M$ be a bilinear map such that for any bilinear map $T : V \times W \to U$ there is a unique $h : M \to U$ with $T = h \circ j$, then M is isomorphic to $V \otimes W$.*

Proof Let $T = i$ and $U = V \otimes W$. Then, $h : M \to V \otimes W$. Conversely, from Definition (A.1), we have $g : V \otimes W \to M$. These satisfy $i = h \circ j$ and $j = g \circ i$. It is easy to check that the uniqueness of h and g implies $hg = 1, gh = 1$ and they are isomorphisms between $V \otimes W$ and M.

THEOREM A1.3 *Tensor products exist.*

Proof Let M be the free vector space on $V \times W$. That is, M is an infinite dimensional vector space with the elements of $V \times W$ as a basis. Let N be the subspace of M generated by elements of the form

$$\begin{aligned} &\text{i)} \quad (v, w_1 + w_2) - (v, w_1) - (v, w_2), \\ &\text{ii)} \quad (v_1 + v_2, w) - (v_1, w) - (v_2, w), \\ &\text{iii)} \quad (rv, w) - r(vw), \\ &\text{iv)} \quad (v, rw) - r(v, w). \end{aligned} \tag{A1.2}$$

Let i be the composition

$$i : V \times W \to M \to M/N, \qquad (A1.3)$$

where the first map is the inclusion and the second map is the natural projection. Then, since the elements of M given in (A1.2) go to zero under the natural map $\rho : M \to M/N$, we see that i is bilinear. Let $T : V \times W \to U$ be a bilinear map. Then there is a linear map $\tilde{T} : M \to U$, such that $\tilde{T}|V \times W = T$. Since T is bilinear $\tilde{T} \mid N = 0$, and therefore, there is a map $g : M/N \to U$ such that $\tilde{T} = g \circ \rho$. Thus $T = g \circ i$. Since the image of i generates the whole of M/N we see that g is uniquely determined. Thus $M/N = V \otimes W$ and $i : V \times W \to V \otimes W$ is the tensor product.

It is usual to denote the element $i(v, w)$ by $v \otimes w$. The space $V \otimes W$ is then called the tensor product of V and W and all reference to i is usually supressed. In the case when V and W are finite dimensional, we can give a more explicit description of $V \otimes W$. Since, in our applications, we only use finite dimensional vector spaces, we now restrict ourselves to this case.

THEOREM A1.4   *Let $v_1, \ldots, v_m$ be a basis for V and $w_1, \ldots, w_n$ be a basis for W. Then, $\{v_i \otimes w_j : 1 \leq i \leq m, 1 \leq j \leq n\}$ is a basis for $V \otimes W$.*

Proof   First, we show that $\{v_i \otimes w_j\}$ is linearly independent. Define a bilinear map T_{ij} on $V \times W$ by

$$T_{ij}\left(a_1 v_1 + \ldots + a_m v_m, b_1 w_1 + \ldots + b_n w_n\right) = a_i b_j. \qquad (A1.4)$$

Let g_{ij} be the map associated to T_{ij}. Then

$$g_{ij}\left(\alpha_{11} v_1 \otimes w_1 + \ldots + \alpha_{mn} v_m \otimes w_n\right) = \alpha_{ij}. \qquad (A1.5)$$

If $\sum \alpha_{ij} v_i \otimes w_j = 0$, and g_{ij} is applied, each coefficient α_{ij} must be zero, which proves linear independence.

The image of $i : V \times W \to V \otimes W$ generates $V \otimes W$. Since the image of i is contained in the span of $\{v_i \otimes w_j\}$, we see that $\{v_i \otimes w_j\}$ spans $V \otimes W$, and therefore, with the first part, this is a basis of $V \otimes W$.

We now introduce the dual space V^* of a vector space V:

$$V^* = \mathrm{Hom}(V, F). \qquad (A1.6)$$

In equation (A1.6) F is the field of scalars. In this work, F is, usually, either the real or the complex numbers.

Theorem A1.5 *Let U, V and W be three finite dimensional vector spaces. Then, the following are isomorphic:*

$$\text{i)} \quad V \otimes W \cong W \otimes V,$$
$$\text{ii)} \quad (U \otimes V) \otimes W \cong U \otimes (V \otimes W),$$
$$\text{iii)} \quad V^* \otimes W \cong \text{Hom}(V, W).$$

Proof Using Theorem A1.4, we can write the isomorphisms as follows:

$$\text{i)} \quad v \otimes w \to w \otimes v,$$
$$\text{ii)} \quad (u \otimes v) \otimes w \to u \otimes (v \otimes w),$$
$$\text{iii)} \quad v^* \otimes w \to (v \to v^*(v)w). \tag{A1.7}$$

To see that iii) is an isomorphism, we need to use the dual basis $v_1^*, \ldots, v_m^*$ to $v_1, \ldots, v_m$ given by

$$v_i^*(v_j) = \begin{cases} 1 & \text{if } i = j, \\ 0 & \text{if } i \neq j. \end{cases} \tag{A1.8}$$

Then, if $T \in \text{hom}(V, W)$, $T(v) = \Sigma \alpha_{ij} v_i^*(v) w_j$.

Invariant proofs, that is, coordinate free proofs of Theorem A1.5 can be given. They have the advantage of showing that the isomorphisms given in (A1.7) are natural. The proofs also show how this theorem generalizes to the infinite dimensional case. However, they require more sophisticated work than is necessary in this book.

The point of part ii) of Theorem A1.5, is to allow us to drop the brackets and write the triple tensor product as $U \otimes V \otimes W$. Thus, we can unambiguously define $\otimes^n V$ by the recurrence relation

$$\otimes^2 V = V \otimes V,$$
$$\otimes^n V = (\otimes^{n-1} V) \otimes V. \tag{A1.9}$$

Then, this allows us to define the tensor algebra

$$T(V) = F \oplus V \oplus (\otimes^2 V) \oplus (\otimes^3 V) \oplus \ldots . \tag{A1.10}$$

This is used in the construction of both the universal enveloping algebra and the Clifford algebra.

Appendix 2: Clifford Algebras and the Spin Groups

To describe the groups $Spin(n)$ it is convenient to start by constructing the Clifford algebra $\mathrm{Cliff}(n)$. Let $V = \mathbf{R}^n$ and $< X, Y >$ be the usual innerproduct on V. Then, as in (10.1), we construct the tensor algebra of V:

$$T(V) = \mathbf{R} \oplus V \oplus (V \otimes V) \oplus \cdots. \qquad (A2.1)$$

Inside $T(V)$ there is the ideal $I(V)$ generated by the following terms:

$$X \otimes Y + Y \otimes X + 2 < X, Y > 1. \qquad (A2.2)$$

DEFINITION A2.1   The Clifford algebra is the quotient algebra $\mathrm{Cliff}(n)$

$= T(V)/I(V)$.

The construction of the Clifford algebra is similar to that of the universal enveloping algebra. Furthermore, we note that there is another ideal in $T(V)$: $J(V)$ generated by

$$X \otimes Y + Y \otimes X. \qquad (A2.3)$$

The quotient algebra $T(V)/J(V) = \wedge(V)$ the exterior algebra of V. Thus, the Clifford algebra is somewhat like the exterior algebra and, in fact, as vector spaces (but **not** as algebras) $\mathrm{Cliff}(n) \cong \wedge(V)$.

Let $E_1, \ldots, E_n$ be the standard basis for $\mathbf{R}^n$. Then, since $V \subset T(V)$, we can regard $E_1, \ldots, E_n$ as elements of $\mathrm{Cliff}(n)$. Using Clifford multiplication, we see these satisfy the relations:

$$E_i E_j = -E_j E_i \quad (i \neq j)$$

$$\text{and} \quad E_i^2 = -1. \qquad (A2.4)$$

An alternative approach to the definition of $\mathrm{Cliff}(n)$ is to define it as "the algebra generated by $E_1, \ldots, E_n$ subject to the relations (A2.4)."

The groups $Spin(n)$ are certain multiplicative subgroups of $\mathrm{Cliff}(n)$. To be a multiplicative group $Spin(n)$ must be contained in the group of units of $\mathrm{Cliff}(n)$, that is, in the group of elements which have a multiplicative inverse. For $n \geq 3$ not every non-zero element of $\mathrm{Cliff}(n)$ has a multiplicative inverse as we see from the next result.

PROPOSITION A2.2   *For $n \geq 3$ the algebra $\mathrm{Cliff}(n)$ has zero divisors.*

Proof   We calculate:

$$(1 + E_1 E_2 E_3)(1 - E_1 E_2 E_3)$$

$$= 1 - E_1 E_2 E_3 E_1 E_2 E_3 = 0. \qquad (A2.5)$$

Definition A2.3   Let $C(n)^*$ be the group of units, that is, invertible elements in $\mathrm{Cliff}(n)$.

The next problem is to find some elements in $C(n)^*$. To do this, we use the injection $i : V \to \mathrm{Cliff}(n)$. Let S^{n-1} be the set of unit vectors in V.

Proposition A2.4   *The elements $i\left(S^{n-1}\right)$ are invertible.*

Proof   Let $X \in S^{n-1}$.

Then,

$$X^{-1} = -X. \qquad (A2.6)$$

We see this since in $\mathrm{Cliff}(n)$ we have $XY + YX = -2 < X, Y >$. Thus $X^2 = - < X, X >$. Hence,

$$X(-X) = -X^2 = < X, X > = 1. \qquad (A2.7)$$

Definition A2.5   The group $Pin(n)$ is the subgroup of $C(n)^*$ generated by $i\left(S^{n-1}\right)$.

The map $X \to -X$ of V, which appeared in (A2.6), is the key to studying the group $Pin(n)$. First, we must extend this to $\mathrm{Cliff}(n)$. It extends in two different ways.

Definition A2.6   The canonical automorphism $\alpha : \mathrm{Cliff}(n) \to \mathrm{Cliff}(n)$ is the extension of the negative of the injection, that is of $x \to -i(x)$, as an automorphism.

The other extension is called conjugation. First, we define the transposition map on $T(V)$ by

$$(X_1 \otimes X_2 \otimes \ldots X_r)^t = X_r \otimes X_{r-1} \otimes \cdots \otimes X_1. \qquad (A2.8)$$

Since $I(V)$ is invariant under transposition, this induces a transposition of $\mathrm{Cliff}(n)$.

Definition A2.7   Conjugation in $\mathrm{Cliff}(n)$ is $\bar{u} = \alpha\left(u^t\right)$.

Notice that conjugation is the extension of $-i$ as an anti-automorphism, that is,

$$\overline{(E_i E_j)} = -E_i E_j$$

$$\text{while} \quad \alpha\,(E_i E_j) = E_i E_j. \tag{A2.9}$$

Both conjugation and α are multiplication by -1 on $i(V)$.

We define a map $\rho : \text{Cliff}(n) \to Aut(\text{Cliff}(n))$ by

$$\rho(u)v = \alpha(u)v\bar{u}. \tag{A2.10}$$

If $v = Y \in V$ there is no reason why $\rho(u)Y$ should be an element of V. Sometimes, however, it is.

DEFINITION A2.8   The Clifford group Γ_n is the set of $u \in C(n)^*$ such that $\rho(u)Y \in \mathbf{R}^n$ for all $Y \in \mathbf{R}^n$.

Clearly, since α is an automorphism and conjugation is an anti-automorphism ρ is a homomorphism form Γ_n into $Aut(\mathbf{R}^n)$.

THEOREM A2.9   a) $Pin(n) \subset \Gamma_n$,
                     b) $\rho : Pin(n) \to O(n)$ *and* $\ker \rho = \{1, -1\}$.

Proof   In $\text{Cliff}(n)$ we have for $X, Y \in \mathbf{R}^n$,

$$YX = -XY - 2 < X, Y > . \tag{A2.11}$$

Thus, for $X \in S^{n-1}$ and $Y \in \mathbf{R}^n$ we have

$$\begin{aligned}
\rho(X)Y &= (-X)Y(-X) \\
&= XYX \\
&= -X^2Y - 2 < X, Y > X \\
&= Y - 2 < X, Y > X. \tag{A2.12}
\end{aligned}$$

Thus, for $X \in S^{n-1}$, $\rho(X)$ is a reflection in the plane perpendicular to X. Since S^{n-1} generates $Pin(n)$ we have proved part a) and that ρ maps $Pin(n)$ into $O(n)$. Furthermore, the image $\rho(Pin(n))$ is generated by reflections. The next stage is to make sure reflections generate $O(n)$. First, notice that in $\mathbf{R}^2$ reflection $r(\theta)$ in the line perpendicular to $X = (\cos\theta/2, \sin\theta/2)$ is given by the matrix

$$r(\theta) = \begin{pmatrix} -\cos\theta & -\sin\theta \\ -\sin\theta & \cos\theta \end{pmatrix}. \tag{A2.13}$$

Thus, reflections generate $O(2)$. By working in each invariant plane in turn, we see from Example 6.21 that reflections generate the maximal torus of $O(n)$. Let $A \in O(n)$. Then, there is a B such that

$$BAB^{-1} \in T. \tag{A2.14}$$

Thus, there is a sequence of reflections $R_1, \ldots, R_k$ so that

$$BAB^{-1} = R_1 \cdots R_k. \tag{A2.15}$$

Hence

$$A = \left(B^{-1}R_1B\right)\left(B^{-1}R_2B\right)\cdots\left(B^{-1}R_kB\right). \tag{A2.16}$$

If R is a reflection in a plane perpendicular to X, then $B^{-1}RB$ is a reflection in a plane perpendicular to $B^{-1}X$. This proves that ρ maps $Pin(n)$ onto $O(n)$.

It only remains to show $\ker\rho = \{1, -1\}$. Clearly, since $\rho(X) = \rho(-X)$ for $X \in S^{n-1}$, 1 and -1 are both in $\ker\rho$. If $X \in S^{n-1}$, we have that $X \notin \ker\rho$, as shown in (A2.12). Suppose $E_{i_1} \ldots E_{i_r} \in \ker\rho$. Then,

$$E_{i_1} \ldots E_{i_r} X E_{i_r} \ldots E_{i_1} = X \tag{A2.17}$$

for all $X \in \mathbb{R}^n$. In particular, if we take $X = E_{i_1}$, we find

$$
\begin{aligned}
E_{i_1} &= E_{i_1} \cdots E_{i_r} E_{i_1} E_{i_r} \cdots E_{i_1} \\
&= (-1)^{r-1} E_{i_1} E_{i_1} \cdots E_{i_r} E_{i_r} \cdots E_{i_1} \\
&= (-1)^{2r-1} E_{i_1} = -E_{i_1}. \tag{A2.18}
\end{aligned}
$$

This contradiction shows that $E_{i_1} \cdots E_{i_r} \notin \ker\rho$ for $r \geq 1$. Thus, $\ker\rho = \{1, -1\}$.

COROLLARY A2.10   *$Pin(n)$ is the double cover of $O(n)$.*

DEFINITION A2.11   *$Spin(n) = \rho^{-1}SO(n)$.*

COROLLARY A2.12   *$Spin(n)$ is the double cover of $SO(n)$.*

To show that the double cover $Spin(n) \to SO(n)$ is non-trivial, we need to join $+1$ to -1 by an arc in $Spin(n)$. Such an arc is

$$\gamma(t) = \cos t + \sin t\, E_1E_2, \ 0 \leq t \leq \pi. \tag{A2.19}$$

Thus, the fundamental group $\pi_1(Spin(n))$ has index 2 in $\pi_1(SO(n))$. It is a fact that, for $n \geq 3$, $\pi_1(SO(n)) = \mathbb{Z}_2$, and therefore $Spin(n), n \geq 3$, is simply connected.

Solutions and Hints for Selected Exercises

Chapter 1

1.1 Consider the map $M(n) \to \mathbf{R}^{n^2}$ by $(a_{ij}) \mapsto (X_k)$ with $X_k = a_{ij}$ for $k = n(i-1) + j$.

1.2 Use Definition 1.9 to compute $[fX, gY](p)(h)$ for any point p and function h.

1.3 - 1.5 Use Theorem 1.11 to compute $[X, X]\tilde{\ }$ and Lemma 1.10 to finish the question. This approach will also work for Exercises 1.4 and 1.5.

1.6 - 1.8 These questions require a routine check of the properties in Definition 1.5.

1.9 First, identify a tangent vector $X \in T_p(M)$ with $a\frac{d}{dx}$ by $X(f) = a\frac{df}{dx}$. Now, apply Definition 1.15.

1.10 Define a map $F : \mathbf{R}^n \to T_p(M)$ by $F(v)(f) = <\text{grad } f, v>$. Now, show that $F(rv) = rF(v)$ for $r \in \mathbf{R}$ and $F(u+v) = F(u)+F(v)$; thus F is a linear map. Next, show that $F(v) = 0$ implies $v = 0$: so F is injective. Finally, observe $\dim \mathbf{R}^n = \dim T_p(M) = n$, and therefore, F is an isomorphism. This uses the general result established at (1.3). Alternatively, one could repeat the argument given there in this special case.

Chapter 2

2.1 First, carry out the check that $SO(3)$ is a group, that is, check if $A, B \in SO(3)$, then $A^{-1} \in SO(3)$ and $AB \in SO(3)$. To show that $SO(3)$ is a manifold, you can either construct an explicit atlas as we did in Example 2.14, or embed $SO(3) \to M(3) \to \mathbf{R}^9$ and use the implicit function theorem to see that $AA^t = I$ and $\det A = 1$ give $SO(3)$ as the inverse image of a point for a suitable map $\mathbf{R}^9 \to \mathbf{R}^{10}$.

2.2 This is a straight-forward computation.

2.3 - 2.4 Use the facts that $\det(A^t) = \det A$ and $\det(\overline{A}) = \overline{\det(A)}$.

2.4 - 2.7   This starts as a routine computation. Then, solve some simple equations. Finally, use w, z and θ to define coordinates on $U(2)$.

2.8 - 2.9   First, observe that if $A \in O(3)$, then A has either one real and two complex or three real eigenvalues. Thus, observe that each eigenvalue has absolute value one. If there are complex eigenvalues, these are complex conjugates. Finally, note that the product of the eigenvalues is $+1$ in 2.8 and -1 in 2.9.

2.10   This is analogous to the Exercises 2.8 and 2.9.

Chapter 3

3.1 $\exp \begin{pmatrix} ix & 0 \\ 0 & -ix \end{pmatrix} = \begin{pmatrix} e^{ix} & 0 \\ 0 & e^{-ix} \end{pmatrix}.$

3.2 $\exp \begin{pmatrix} 0 & z \\ -\bar{z} & 0 \end{pmatrix} = \begin{pmatrix} \cos|z| & \frac{z}{|z|}\sin|z| \\ \frac{-\bar{z}}{|z|}\sin|z| & \cos|z| \end{pmatrix}.$

3.3   Check $f(t_1 + t_2) = f(t_1)f(t_2)$.

3.5   Find the eigenvalues and eigenvectors of $\begin{pmatrix} ix & z \\ -\bar{z} & -ix \end{pmatrix}$.

3.6   Use the formula $\exp(PAP^{-1}) = P(\exp A)P^{-1}$.

3.7   Check $f^g(t_1 + t_2) = f^g(t_1)f^g(t_2)$.

3.8   Use formula (3.7).

3.9   Check that gAg^{-1} is conjugate skew symmetric and $tr(gAg^{-1}) = 0$.

3.10   Let $f_v : \mathbf{R} \to H$ be the one parameter subgroup of $v \in \mathcal{H}$. Then, show $i \circ f_v : \mathbf{R} \to G$ where $i : H \to G$ is the inclusion map is the one parameter subgroup of $v \in \mathcal{G}$.

Chapter 4

4.1   This exercise is easy and straightforward, but long. The answer is:

$$Z_1 \begin{pmatrix} ix & z \\ -\bar{z} & -ix \end{pmatrix}, \quad Z_2 = \begin{pmatrix} 0 & i\bar{x}z \\ ix\bar{z} & 0 \end{pmatrix}, \quad Z_3 = \frac{1}{6}\begin{pmatrix} -ixz\bar{z} & -x^2z \\ x^2\bar{z} & ixz\bar{z} \end{pmatrix}.$$

4.3 Check this using the series (3.32).

4.4 Use the facts that $\det A^t = \det A$ and for an $n \times n$ matrix $\det(-A) = (-1)^n \det A$. Now $A \in SO(2l+1)$ has a zero eigenvalue, and hence, has the given form. The last statement follows from the formula $\exp(PBP^{-1}) = P(\exp B)P^{-1}$.

4.5 $W_1 = 0$, $W_2 = -\frac{1}{2}[X,Y]$, $W_3 = -\frac{1}{3}[[X,Y],Y] - \frac{1}{6}[[X,Y],X]$.

4.6 This is a long computation. It is recommended that you try the special cases $x = 0$ and $z = 0$ first. If $\begin{pmatrix} \alpha & \beta \\ -\bar{\beta} & \bar{\alpha} \end{pmatrix} \in SU(2)$ then $\begin{pmatrix} \alpha & \beta \\ -\bar{\beta} & \bar{\alpha} \end{pmatrix} = P\begin{pmatrix} e^{i\theta} & 0 \\ 0 & e^{-i\theta} \end{pmatrix}P^{-1}$, and therefore, $\begin{pmatrix} \alpha & \beta \\ -\bar{\beta} & \bar{\alpha} \end{pmatrix}^n = P\begin{pmatrix} e^{in\theta} & 0 \\ 0 & e^{-in\theta} \end{pmatrix}P^{-1}$. The difficulties in carrying this out explicitly illustrate the advantages of using the Campbell-Baker-Hausdorff formula.

4.8 The map $S^1 \times \ldots \times S^1 \to T$ by $(e^{i\theta_1}, \ldots, e^{i\theta_{n-1}}) \mapsto (e^{i\theta_1}, \ldots, e^{i\theta_{n-1}}, e^{i\theta_n})$ with $\theta_n = -\sum_{j=1}^{n-1}\theta_j$ shows that T is a torus.

4.9 First, observe $\exp_T : T \to T$ is onto and then, by Exercise 3.10, $\exp_T = \exp_G |T$.

4.10 The map $x \mapsto gxg^{-1}$ is an isomorphism between T and gTg^{-1}.

Chapter 5

5.1 Use the formula $ad(A)X = AX - XA$.

5.3 The map $adX : \mathcal{G} \to \mathcal{G}$ is the zero map, and therefore, its matrix relative to any basis is the zero matrix.

5.4 With X in the center of $\mathcal{G}$, the hint of Exercise 5.3 applies again.

5.5   First, observe that if $X = i\,\theta I$, then $tr\,adX\,adY = 0$ for all Y since adX is the zero map. Next, let Y be the diagonal matrix with entry i in the jth place. Then, if X has entries $i\,\theta_j$ in the jth place, we have $-2n\theta_j + 2\sum_j \theta_j = 0$; thus $\theta_j = \frac{1}{n}\sum\theta_j$ and all diagonal entries are equal.

5.6   Let T denote the set of diagonal matrices in $U(n)$ with Lie algebra $\mathcal{T}$. Then $\exp : \mathcal{T} \to T$ is onto and, by Exercise 5.4, if $A \in T$ is in the center of $U(n)$ then $\exp^{-1}(A)$ is in the kernel of the Killing form. Thus, by Exercise 5.5, $\exp^{-1}(A) = i\,\theta I$ and $A = e^{i\,\theta}I$.

5.7   Since G is semisimple, the kernel of the Killing form is just zero, and therefore, by Exercise 5.4 the center Z is discrete. Since G is compact, any discrete subgroup is finite.

5.9   Clearly, $Z \subset \text{Ker }Ad$. Conversely, let $X \in \text{Ker }Ad$. Then $(dI_x)_1 = 0$. Now, $I_x L_g = L_{xgx^{-1}}I_x$, and therefore, $d(I_x)_g = 0$. Pick a neighborhood U of 1 in the group and then use local coordinates to show I_x is the identity on U, that is, $I_x g = g$ for $g \in U$. Since U generates the group $I_x g = g$ for all elements of the group, and therefore, $X \in Z$.

Chapter 6

6.1   Suppose $z \in Z$ and $z \notin T$. Let H be the subgroup of G generated by T and z. Then, show H is abelian and contains T, which contradicts the maximality of T. Alternatively, note there is y such that $yzy^{-1} \in T$. Hence $yzy^{-1} \neq z$ but $z \in Z$.

6.3   By Exercise 6.1, the center of $U(n)$ is contained in the set of diagonal matrices, and therefore, Exercise 5.6 applies to yield the result.

6.4   Observe that $SU(n) \subset U(n)$ and center of $SU(n) \subset$ center of $U(n)$. Now, by Exercise 6.1, center of $SU(n) \subset$ maximal torus of $SU(n)$. Thus, since $\mathbb{z}_n =$ maximal torus of $SU(n) \cap$ center of $U(n)$, we have center of $SU(n) \subset \mathbb{z}_n$. The reverse inclusion is easy.

6.7   Use the same idea as outlined in Exercise 6.6.

6.8   The method to find $\tilde{q}$ is described in the proof of Lemma 6.28: $\tilde{q} = a + ib + jc + kd$ where

$$a + ib = \frac{-w}{2 - (e^{i\theta}\bar{z} + e^{-i\theta}z)} \quad \text{and} \quad c + id = \frac{-w}{2 - (e^{i\theta}z + e^{-i\theta}\bar{z})}.$$

Notice that in both expressions the denominators are real, therefore, a, b, c, d can be found by dividing the real and imaginary parts of $-w$ by the appropriate denominator.

6.9 A routine computation shows $\overline{\chi(A)}^t = \chi\left(\bar{A}^t\right)$ and $\sigma(Av) = \chi(A)\sigma(v)$. Thus $\chi(A) \in U(2n)$. Now, notice that this is sufficient to prove Theorem 6.32 and use this to show $\det \chi(A)$ is a positive real number. Since $\chi(A) \in U(2n)$ we know $\det \chi(A)$ has absolute value 1 and so $\det \chi(A) = 1$. Thus $\chi(A) \in SU(2n)$.

6.10 This is a routine computation.

Chapter 7

7.1 - 7.2 Check the conditions of the definition of a representation.

7.3 The space V_1 is one-dimensional, and therefore, it is irreducible. To see V_2 is irreducible observe that $\dim V_2 = 3$ so if V_2 has a non-trivial invariant subspace, it must have a one-dimensional invariant subspace. That is, there must be an eigenvector of $\pi(x)$ which is independent of x. First, calculate the eigenvectors of $\pi \begin{pmatrix} e^{i\theta} & 0 \\ 0 & e^{-i\theta} \end{pmatrix}$. Then, observe that none of these are eigenvectors of $\pi \begin{pmatrix} 1/\sqrt{2} & 1/\sqrt{2} \\ -1/\sqrt{2} & 1/\sqrt{2} \end{pmatrix}$.

7.4 $\operatorname{tr} \pi_1 \begin{pmatrix} z & w \\ -\bar{w} & \bar{z} \end{pmatrix} = 1$, $\operatorname{tr} \pi_2 \begin{pmatrix} z & w \\ -\bar{w} & \bar{z} \end{pmatrix} = (z + \bar{z})^2 - 1$.

7.5 Notice that if $\begin{pmatrix} z & w \\ -\bar{w} & \bar{z} \end{pmatrix} = g \begin{pmatrix} e^{i\theta} & 0 \\ 0 & e^{-i\theta} \end{pmatrix} g^{-1}$, then $\chi_n \begin{pmatrix} z & w \\ -\bar{w} & \bar{z} \end{pmatrix} = \chi_n \begin{pmatrix} e^{i\theta} & 0 \\ 0 & e^{-i\theta} \end{pmatrix}$ and use formula (7.21) .

7.7 Let S and T be linear transformations of V_m and V_n. Let V_m have a basis $u_1, \ldots, u_{m+1}$ and V_n have a basis $v_1, \ldots, v_{n+1}$. Then $\{u_i \otimes v_j\}$ is a basis for $V_m \otimes V_n$. If S has matrix (s_{ij}) so $Su_i = \sum_j s_{ij} u_j$ and T has matrix (t_{ij}) then $S \otimes T(u_i \otimes v_j) = \sum_{\kappa, \ell} s_{i\kappa} t_{j\ell} u_\kappa \otimes v_\ell$. Thus, we have $\operatorname{tr}(S \otimes T) = \sum_{i,j} s_{ii} t_{jj} = (\sum_i S_{ii})(\sum_j t_{jj}) = \operatorname{tr} S \operatorname{tr} T$. In particular, taking $S = \pi_m(g)$ and $T = \pi_n(g)$ yields the result of this exercise.

7.10 The representation of Exercise 7.1 is $\pi_0 \oplus \pi_2$ as shown by Exercises 7.1, 7.2, and 7.3.

Chapter 8

8.1   See Example 8.24.

8.4   See Example 8.29.

8.6   If we restrict π_λ and π_μ to the maximal torus T, we find π_λ is
made up of blocks $\begin{pmatrix} \cos 2\pi\alpha & -\sin 2\pi\alpha \\ \sin 2\pi\alpha & \cos 2\pi\alpha \end{pmatrix}$ and π_μ of blocks
$\begin{pmatrix} \cos 2\pi\beta & -\sin 2\pi\beta \\ \sin 2\pi\beta & \cos 2\pi\beta \end{pmatrix}$. Over the complex numbers, these can be di-
agonalized as $\begin{pmatrix} e^{2\pi i\alpha} & 0 \\ 0 & e^{-2\pi i\alpha} \end{pmatrix}$ and $\begin{pmatrix} e^{2\pi i\beta} & 0 \\ 0 & e^{-2\pi i\beta} \end{pmatrix}$. The tensor product of these numbers is $\begin{pmatrix} e^{2\pi i(\alpha+\beta)} & & & \\ & e^{2\pi i(\alpha-\beta)} & & \\ & & e^{-2\pi i(\alpha-\beta)} & \\ & & & e^{-2\pi i(\alpha+\beta)} \end{pmatrix}$.

Thus, the weights of $\pi_\lambda \otimes \pi_\mu$ are $\pm\alpha \pm \beta$. Since, if α is a weight, so is
$-\alpha$, we have the result of the exercise.

8.7   This follows by carefully counting dimensions in Exercise 8.6.

8.9   From the formula (7.20), the weight spaces are spanned by
$x^{n-r}y^r$.

Chapter 9

9.1 $\dim V_\lambda = \frac{1}{2}(m + 1)(n + 1)(m + n + 2)$.

9.2   Use Definition 9.8 to find:

$$\begin{aligned}
j_\lambda &= \exp i((m + n + 2)\theta_1 + (m + 1)\theta_2) - \exp i((m + 1)\theta_1 \\
&\quad + (m + n + 2)\theta_2) \\
&\quad + \exp i(-(m + 1)\theta_1 + (n + 1)\theta_2) - \exp i((n + 1)\theta_1 - (m + 1)\theta_2) \\
&\quad + \exp i(-(n + 1)\theta_1 - (m + n + 2)\theta_2) - \exp i(-(m + n + 2)\theta_1 \\
&\quad - (n + 1)\theta_2),
\end{aligned}$$

and hence,
$$j = \exp i(2\theta_1 + \theta_2) - \exp i(\theta_1 + 2\theta_2)$$
$$+ \exp i(-\theta_1 + \theta_2) - \exp i(\theta_1 - \theta_2)$$
$$+ \exp i(-\theta_1 - 2\theta_2) - \exp(-2\theta_1 - \theta_2).$$

The character is now given by $\chi_\lambda = j_\lambda/j$.

9.3 $\dim V_\lambda = \frac{1}{6}(2m + n + 3)(m + n + 2)(m + 1)(n + 1)$.

9.4 Use Definition 9.7 as in Exercise 9.2.

9.5 Now $j_\lambda(t) = \Sigma(-1)^\sigma \exp(2\pi i\sigma(\lambda + \rho)t)$ so $j_\lambda(-t) = \overline{j_\lambda(t)}$. If $g = \exp t$ then $g^{-1} = \exp(-t)$ and therefore, $\chi_\lambda(g^{-1}) = j_\lambda(-t)/j(-t) = \overline{j_\lambda(t)}/\overline{j(t)} = \overline{\chi_\lambda(g)}$.

9.6 Let $N(T)$ be the normalizer of T in G. If $\sigma \in W$ there is $s \in N(T)$ so that $\sigma(t) = sts^{-1}$.

9.8 Compute $\int_G f(sxs^{-1}y)g(y^{-1})dy$ using the change of variables $z = s^{-1}ys$ and the invariance of the measure.

Chapter 10

10.1, 10.2, 10.4, 10.5 These are all different applications of the formula (10.23).

10.3 If we identify $C^\infty(G)$ as a representation space of G, then the action of the Laplacian on $C^\infty(G)$ corresponds to the action of the Casimir element on the representation space. Since the Casimir acts as a scalar on each irreducible representation space, the result follows.

Chapter 11

11.1 Use formula (11.7) to calculate $\frac{1}{2}(X_ig_{jk} + X_jg_{ik} - X_kg_{ij})$, remembering $\nabla_{X_i}X_j = \nabla_{X_j}X_i$.

11.2 Since the connection has zero torsion, we have: $\nabla_X Y - \nabla_Y X = [X, Y]$. Thus, we obtain $< \nabla_X Y, Z > - < \nabla_Y X, Z > = < Z, [X, Y] >$.

Now, $< Y, Z > = 0$, so $X < Y, Z > = < \nabla_X Y, Z > + < Y, \nabla_X Z > = 0$. Thus, if we add the equations $< \nabla_Z Y, X > - < \nabla_Y Z, X > = - < X, [Y, Z] >$ and $< \nabla_Z X, Y > - < \nabla_X Z, Y > = < Y, [Z, X] >$ to the similar one above, we obtain the formula of this exercise.

11.3   Use $< [Z, X], Y > = - < X, [Z, Y] >$.

11.4   The relevant dimensions are: $\dim SU(2) = 3$, $\dim SU(3) = 8$, $\dim SO(3) = 3$, $\dim SO(4) = 6$ and $\dim Sp(2) = 10$.

Chapter 12

12.1, 12.2   These are routine applications of advanced calculus.

12.3   The dimensions of these groups are given in the note on Exercise 11.4.

12.4   See my paper: "The Heat Equation and Modular Forms," Journal of Differential Geometry, **13**(1978), 589–602.

Bibliography

[1] J. F. Adams, *Lectures on Lie Groups*, Benjamin, New York, 1969.

[2] M. F. Atiyah, et al., *Representation Theory of Lie Groups*, Proceedings of the SRC/LMS Research Symposium on Representations of Lie Groups, Oxford 28th June - 15 July, 1977, Cambridge University Press, Cambridge, 1979.

[3] M. L. Curtis, *Matrix Groups*, Springer, New York, 1984.

[4] J. Dieudonné, *Treatise on Analysis*, Vol. V, Academic Press, New York, 1977.

[5] H. Freudenthal and H. de Vries, *Linear Lie Groups*, Academic Press, New York, 1969.

[6] S. Helgason, *Differential Geometry, Lie Groups and Symmetric Spaces*, Academic Press, New York, 1978.

[7] J. E. Humphreys, *Introduction to Lie Algebras and Representation Theory*, Springer, New York, 1972.

[8] N. Jacobson, *Lie Algebras*, Dover Publications, New York, 1979.

[9] J. Milnor, *Morse Theory*, Annals of Math. Studies, No. 51, Princeton University Press, Princeton, NJ, 1969.

[10] J. F. Price, *Lie Groups and Compact Groups*, Cambridge University Press, Cambridge, 1977.

Index